Developments in expert systems

Computers and People Series

Edited by
B. R. GAINES

The series is concerned with all aspects of man–computer relationships, including interaction, interfacing modelling and artificial intelligence. Books are interdisciplinary, communicating results derived in one area of study to workers in another. Applied, experimental, theoretical and tutorial studies are included.

Developments in expert systems

From a special issue of the
International Journal of Man–Machine Studies

edited by

M. J. COOMBS

Department of Computer Science
University of Strathclyde
Glasgow G1 1HX
Scotland

1984

ACADEMIC PRESS
(A Subsidiary of Harcourt Brace Jovanovich, Publishers)

London   Orlando   San Diego   San Francisco   New York
Toronto   Montreal   Sydney   Tokyo   Toronto

ACADEMIC PRESS INC. (LONDON) LTD
24/28 Oval Road, London NW1 7DX

United States Edition published by
ACADEMIC PRESS INC.
111 Fifth Avenue, New York, New York, 10003

British Library Cataloguing in Publication Data

Developments in expert systems.
1. Expert systems (Computer science)
I. Coombs, M. J.
001.64   QA76.9.E96

ISBN 0-12-187580-6

Filmset and printed in UK by J. W. Arrowsmith Ltd., Bristol

may 85

Contributors

J. Alty
Department of Computer Science
University of Strathclyde
Glasgow G1 1XH
UK

A. Basden
Decision Support Systems Group
Corporate Manage Services
I.C.I. plc
P.O. Box 11, The Heath
Runcorn, Cheshire
UK

B. Chandrasekaren
Department of Computer and
    Information Science
The Ohio State University
Columbus
Ohio 43210
USA

W. J. Clancey
Heuristic Programming Project
Computer Science Department
Stanford University
Stanford
California 943045
USA

M. Coombs
Department of Computer Science
University of Strathclyde
Glasgow G1 1XH
UK

R. Davis
The Artificial Intelligence Laboratory
Massachusetts Institute of Technology
Cambridge
Massachusetts 02139
USA

D. W. Hasling
Heuristic Programming Project
Computer Science Department
Stanford University
Stanford
California 94305
USA

P. Jackson
Institute of Educational Technology
Open University
Milton Keynes
MK7 6AA

L. N. Kenal
University of Maryland
Baltimore
Maryland
USA

J. L. Kolodner
School of Information and Computer
    Science
Georgia Institute of Technology
Atlanta
Georgia 30332
USA

B. A. Lambird
L.N.K. Corporation
Silver Spring
Maryland
USA

C. P. Langlotz
Heuristic Programming Project
Departments of Medicine and Computer
    Science
Stanford University
Stanford
California 94305
USA

D. Lavine
L.N.K. Corporation
Silver Spring
Maryland
USA

P. Lefrère
Institute of Educational Technology
Open University
Milton Keynes
MK7 6AA

S. Mittal
Department of Computer and
  Information Science
The Ohio State University
Columbus
Ohio 43210
USA

D. S. Nau
Computer Science and Mathematics
  Department
University of Maryland
Baltimore
Maryland
USA

J. A. Reggia
Department of Neurology
University of Maryland
Baltimore
Maryland
USA

C. K. Riesbeck
Department of Computer Science
Yale University
New Haven
Connecticut
USA

G. Rennels
Heuristic Programming Project
Computer Science Department
Stanford University
Stanford
California 94305
USA

E. H. Shortliffe
Heuristic Programming Project
Departments of Medicine and Computer
  Science
Stanford University
Stanford
California 94305
USA

P. Y. Wang
Department of Mathematics
University of Maryland
Baltimore
Maryland
USA

G. P. Zarri
Centre national de la Recherche
  Scientifique
Laboratoire d'Informatique pour les
Sciences de l'Homme
54, Boulevard Raspail
75270 Paris Cedex 06
France

A. C. Zimmer
Department of Psychology
Westfälische Wilhelms Universität
Münster
Federal Republic of Germany

Preface

The potential value of automated knowledge-based problem-solvers—or "expert systems"—is now firmly established. Since the building of the first system 15 years ago, several exemplar projects have demonstrated that computer programs are capable of expert performance in diagnostic, interpretive, control and planning tasks within a wide range of domains.

Despite a comfortable number of successful demonstration projects, medicine, industry and commerce have been slow to adopt expert systems for routine use. However, although the reasons for this are complex, a number of factors may be identified. These include the poor "fit" between the tasks undertaken by expert systems in many areas and those performed by a human expert, the opaque reasoning of many systems and the inability of systems to learn from experience. It is issues such as these which are the topics of current research and which are discussed in the chapters of this volume "Developments in Expert Systems".

March, 1984   M. J. Coombs

Contents

ix

Introduction

Mike Coombs

It is now over 15 years since the DENDRAL research group at Stanford demonstrated the potential of the young discipline of Artificial Intelligence to undertake problem-solving in complex, real-world domains. In contrast to much of the work being conducted at the time, which sought principally to develop powerful, general purpose problem-solving methods, the Stanford group took the alternative approach of mechanizing the domain knowledge actually used by experts. Thus, the DENDRAL program successfully achieved its objective of inferring the structure of chemical compounds from mass-spectral data by using explicit knowledge of mass spectrometry (Buchanan and Feigenbaum, 1978).

The incorporation of explicit domain knowledge into problem-solving programs proved to be of great practical and theoretical importance. First, it enabled Artificial Intelligence to solve many real-world problems which were previously beyond its powers. From this initial exploration of "knowledge-programming" a whole tradition has developed, covering such diverse areas of application as teaching systems (see the collection of papers edited by Sleeman and Brown (1982) in the "Computers and People Series"), natural language understanding (for example, Schank and Abelson, 1977) and diagnostic problem-solving (for example, Shortliffe, 1976). Secondly, the knowledge-based approach generated its own problems which had the effect of extending the theoretical interests of the subject, raising not only issues of knowledge representation but also questions of the structure of the different cognitive functions programs were required to simulate (e.g. story understanding, question answering, fault finding).

From an application point of view, the most successful use of knowledge programming has remained with problem-solving in areas that require considerable expertise—where it began with DENDRAL. There are now a substantial number of "expert systems" which can claim expert, or near expert performance in a wide range of domains, undertaking such problem-solving tasks as medical diagnosis, data analysis and planning. Some of the best known systems include:

MYCIN, a system for diagnosing bacterial infections (Shortliffe, 1976);
PROSPECTOR, a system for assisting the evaluation of geological prospects (Duda et al., 1978);
CASNET/glaucoma, a system for diagnosing the eye disease glaucoma (Weiss, Kulikoski and Safir, 1977);

R1, a system for configuring VAX computers (McDermott, 1980);
Internist-1 and CADUCEUS, advisors for general internal medicine (People, 1982) and
MOLGEN, an automated "scientist's assistant" in the field of molecular genetics (Friedland, 1979; Stefik, 1980).

Fifteen years is sufficient period of time for the development of a clear set of general principles to characterize an area. However, within expert systems these principles are far from general and have come to form an orthodoxy, the main dictates of which have been noted by Davis (1982):

separate the inference engine from the knowledge-base;
use as uniform a representation as possible, the preferred form of representation being production rules;
keep the inference engine simple;
provide some form of facility by which the system can explain itsa conclusions to the user and
favour problems that require the use of substantial bodies of empirical associative knowledge over those that may be solved using causal or mathematical knowledge.

The above principles provided a powerful starting point for expert systems research. The rule-based representation, for example, permitted systems to evolve with experience and placed few restrictions on content. The use of a simple control structure allowed basic explanation facilities, by which a system could explain its reasoning to the user, to be developed with some ease, so aiding the modification and debugging of the knowledge-base. However, the constraints imposed by the orthodoxy have recently been found to be over-restrictive by an increasing number of people.

The articles in this volume, "Developments in Expert Systems" represents accounts of the ways a number of researchers have questioned or broken these constraints. In particular, many of the papers focus on two problem areas:

(1) the extension of expert systems to employ forms of knowledge other than empirical associations and the design of architectures to support them and
(2) the design of systems to undertake a wider range of cognitive functions than the largely pre-determined, system directed problem-solving of earlier programs. This includes, for example, the ability to "shadow" the user's problem-solving, interrupting only when possible error is noted, and also the ability to learn automatically from mistakes.

The first three articles focus upon the use of causal knowledge, supported by non-traditional expert systems architectures. Davis, in the opening chapter, discusses his recent work on the design of a system to diagnose faults in digital electronic hardware from structural and causal models of devices. He argues that where such models exist they should be used, particularly as the principled underpinning to a system makes it easier to construct and update. A central interest is the method proposed for processing the models. Instead of employing the probabilistic reasoning of MYCIN or PROSPECTOR, Davis seeks to construct a "fault model" which will account for violations between the expected and actual operation of a device. This technique is termed "discrepancy detection". Moreover, by explicitly characterizing faults as "models of causal interaction", and organizing them in terms of the complexity of interactions implied, Davis proves that it is possible to diagnose complex faults—e.g. bridge faults—which are beyond other methods.

The theme of using causal knowledge for diagnosis is continued in the chapter by Chandrasekaran and Mittal. In contrast to Davis, however, they argue with reference to a medical domain that there is no need for the causal model to be represented explicitly in the system. It is proposed that causal knowledge may be "compiled" into a diagnostic problem-solving structure in such a way as to solve all problems within its scope as effectively as an explicit model. This contention is supported with reference to a system –MDX– in which a disease is identified by a process of step-wise refinement.

A significant feature of the diagnostic system described in the article by Reggia, Nau and Wang is that it is not heuristic, but is built on a firm mathematical base. It is argued that a modification of the set-covering problem not only supports diagnostic reasoning but also essentially follows the pattern of reasoning used by human experts. Moreover, the technique allows for the possibility of there being multiple diseases present, while constraining the diagnosis to the "minimum" set cover (set of diseases) which will explain the symptoms. These are characteristics of human performance not supported by traditional rule-based systems.

A theme which runs through the first three chapters is that there are advantages in building systems to conform to explicit, pre-defined constraints, either in the structure of the knowledge-base or in the processing model. This idea is continued in chapters by Basden, who focuses upon the issues involved in setting up an expert system development program in a multi-national chemical company. The experience of producing a demonstration system on stress-corrosion-cracking using a commercially available PROSPECTOR-like development package suggested that this architecture was well able to support reasoning from a causal representation of the problem. Moreover, this increased the coherence of the knowledge-base, which made the system easier to debug and its reasoning easier to understand. It thus appears that we should distinguish between those limitations of earlier expert system designs which are due to a lack of constraints imposed on the rule-base and those due to limitations in the architecture itself.

Basden lists a wide range of roles for expert systems, including a tool for the refinement of expertise, a training aid and a medium for communicating expertise. All of these will require system/user relationships which differ from the conventional model of high computer control in eliciting evidence and selecting processing goals. Chapters 5 and 6 discuss these issues. Langlotz and Shortliffe describe the experience of adapting a system–ONCOCIN–which was originally intended to assist with the routine care of cancer patients by handling the collection of clinical data and advising on treatment to adopt the different role of analysing and critiquing the physician's own plans. A critique is an explanation of differences between the plan proposed by the expert system and that of user. This adaptation was found necessary because physicians did not always desire advice and frequently needed to override ONCOCIN's unsolicited suggestions because of minor differences. Given the more passive role, the system could be programmed to ignore minor differences, only interrupting when significant deviations were observed. Kolodner addresses the problem of designing an expert system which is able to improve its performance with experience. Her program –SHRINK– operates within the domain of psychiatric diagnosis and treatment, and uses techniques from work on the role of long-term memory in text understanding (Schank, 1979) rather than models of problem-solving. More particularly, the system refines its expertise by identifying similarities and differences between diagnosis "cases", using similarities to develop more general memory structures and differences to extend the explanatory scope of those structures. By being exposed to successive cases, the program develops both more comprehensive and refined diagnosis knowledge and more effective diagnostic processes.

An important theme in expert systems research is that of "explanation". Effective methods for generating explanations of both system queries and conclusions have long been considered an important feature of expert systems. However, it has proved difficult to design systems which address users with different needs at the right level of abstraction and detail. In Chapter 7, Hasling, Clancey and Rennels, first review a number of topics which bear on the quality of an explanation, namely the domain knowledge and procedures to be explained, the constraints imposed on a successful explanation by user understanding, and effective techniques of expression. They then consider in detail problems of generating good explanations of problem-solving strategy. This is discussed in the context the NEOMYCIN system (an instructional relation of the MYCIN diagnostic system), and focusses upon the use of an explicit representation of control knowledge. It is emphasized that to be effective this knowledge must be complete and must be made in an abstract, domain-independent form.

"Explanation" is also the topic of the next chapter, by Coombs and Alty, which considers the role of user and system-generated explanations in helping experts (rather than inexperts) to solve problems falling at the junction of their knowledge and that of some other expert. Coombs and Alty propose a design based on discovery learning, the aim being to support the user's problem-solving by helping him extend and refine his understanding of surrounding concepts. The system –MINDPAD– provides the user with a database of domain knowledge, a set of procedures for its application to the problem, and an intelligent machine-based advisor to judge the user's effectiveness and to advise on strategy. The procedures employ a comparison between system and user explanations both to promote the application of domain knowledge and to expose understanding difficulties. Database PROLOG provides the subject material for the prototype system.

As a person acquires expertise, significant changes might be expected in the structure of his knowledge. Furthermore, an expert system may need to take these into account when providing explanations of evaluating a user's explanations. A study of differences congruent with increasing expertise in economics is reported by Riesbeck. Riesbeck found that such differences can be both identified and modelled. Although both novices and experts employ the same concepts, novices organize their reasoning around the goals of a few factors, while experts represent their knowledge as a few explicit, specific rules directly linking economic quantities. The transition between these two stages may itself be modelled using the notion of failure-driven memory (see Kolodner in Chapter 6). A point to be noted by designers of explanation facilities is that while the knowledge of experts was found to be comprehensible to novices, novices could not necessarily reproduce an expert's reasoning. They will thus need significant system support if they are to learn from explanations. Attention to differences in knowledge structures may make such support more effective.

Jackson and Lefrère discuss problems of designing an intelligent machine-based help system for word processing. Arguing that the most effective advisory systems employ some explicit representation of the user's intentions, they seek to define a theoretical context for interpreting user inputs in terms of goals. The framework proposed is derived from recent work on speech acts, planning and metalevel inference. A system would employ this framework to maintain hypotheses concerning the current state of user's plan and to formulate a response to given user query types (i.e. How can I . . .? Do . . .!). Finally, it is proposed that this model may be implemented using techniques developed within rule-based systems research.

In Chapter 11, Zarri describes an application of knowledge processing ideas as to the management of a biographical database. The main techniques are derived from work on natural language understanding, although the spirit of the work follows expert systems research in seeking to simulate the skills of an expert in information retrieval. The system –RESEDA– codes biographical facts from mediaeval French history into frame-like structures using a meta-language based on Case Grammar. One significant advantage of this approach is that the query system may use "deep knowledge" of biographical relations encoded in the meta-language to add new causal information to the database. This will enable historical inferences to be made where these are insufficient facts to give a direct answer.

The article by Lambird, Lavine and Kanal also describes the application of expert system ideas to a new area; that of cartographic feature extraction from remotely-sen sed imagery. They argue that this application requires the integration of support. Finally, they propose a system using a parallel nondirectional search technique and distributed architecture to handle the integration task.

In the final chapter, Zimmer considers the application of expert system ideas to forecasting systems. He points out that traditional, numerically-based computer aids have not been proved able to support the demands upon them and argues that the use of the qualitative systems would make an improvement. These would then be integrated with the numerical system. Zimmer presents a model within the framework of possibility theory which reflects the updating of verbal judgements in the light of new information. This model is proved to follow human performance closely and its applicability is demonstrated with reference to the prediction of exchange rates by bank clerks.

References

BUCHANAN, B. G. & FEIGENBAUM, E. A. (1978). DENDRAL and Meta-DENDRAL: their application dimension. *Journal of Artificial Intelligence*, **11**, 5–24.

DAVIS, R. (1982). Expert systems: Where are we? And where do we go from here? *The A.I. Magazine*, **3**, 3–22.

DUDA, R. O., GASCHNIG, J., HART, P. E., KONOLIGE, K., REBOH, R., BARRETT, P. & SLOCUM, J. (1978). Development of the PROSPECTOR consultation system for mineral exploration. *Final Report, SRI Projects 5821 and 6415*, SRI International, Inc., Menlo Park, California.

FRIEDLAND, P. E. (1979). Knowledge-based experiment design in molecular genetics. *Doctoral dissertation. Report 79-771*, Department of Computer Science, Stanford University, Stanford, California.

MCDERMOTT. J. (1980). R1: a rule-based configurer of computer systems. *Report CMU-CS-80-119*, Department of Computer Science, Carnegie–Mellon University, Pittsburgh, Pennsylvania.

POPLE, H. E. (1982). Heuristic methods for imposing structure on ill-structured problems: the structuring of medical diagnostics. In SZOLOVITS, P., Ed., *Artificial Intelligence in Medicine*. Boulder, Colorado: Westview Press.

SCHANK, R. C. (1979). Reminding and memory organization: an introduction to MOPs. *Report #170*, Department of Computer Science, Yale University.

SCHANK, R. C. & ABELSON, R. P. (1977). *Scripts, Plans, Goals and Understanding.* Hillsdale, New Jersey: Lawrence Erlbaum.

SHORTLIFFE, E. H. (1976). *Computer-based Medical Consultations: MYCIN.* New York: American Elsevier.

STEFIK, M. J. (1980). Planning with constraints. *Doctoral dissertation. Report 80-784*, Department of Computer Science, Stanford University, California.

WEISS, S. M., KULIKOWSKI, C. & SAFIR, A. (1977). A model-based consultation system for the long-term management of glaucoma. *IJCAI—5*, 826–832.

Reasoning from first principles in electronic troubleshooting

RANDALL DAVIS

The Artificial Intelligence Laboratory, Massachusetts Institute of Technology, Cambridge, Massachusetts 02139, U.S.A.

While expert systems have traditionally been built using large collections of rules based on empirical associations, interest has grown recently in the use of systems that reason "from first principles", i.e. from an understanding of causality of the device being examined. Our work explores the use of such models in troubleshooting digital electronics.

In discussing troubleshooting we show why the traditional approach—test generation—solves a different problem and we discuss a number of its practical shortcomings. We consider next the style of debugging known as discrepancy detection and demonstrate why it is a fundamental advance over traditional test generation. Further exploration, however, demonstrates that in its standard form discrepancy detection encounters interesting limits in dealing with commonly known classes of faults. We suggest that the problem arises from a number of interesting implicit assumptions typically made when using the technique.

In discussing how to repair the problems uncovered, we argue for the primacy of *models of causal interaction*, rather than the traditional fault models. We point out the importance of making these models explicit, separated from the troubleshooting mechanism, and retractable in much the same sense that inferences are retracted in current systems. We report on progress to date in implementing this approach and demonstrate the diagnosis of a bridge fault—a traditionally difficult problem—using our approach.

1. Introduction

Many of the early efforts at expert system construction were based on the collection of large numbers of rules that captured empirical associations about their domain. The competence of a program like MYCIN (Shortliffe, 1976), for example, arises from the assembly of several hundred rules that encode the experience of an accomplished diagnostician. We refer to the rules as *empirical associations* because they associate symptoms with diseases, and because they are often justified on purely empirical grounds: in the lifetime of experience of the expert, that symptom/disease pair has been observed often enough that it becomes worth remembering. A number of programs [e.g. INTERNIST (Pople, 1982), PROSPECTOR (Gaschnig, 1979), DIPMETER ADVISOR (Davis *et al.*, 1981)] have been built in this way and have in some cases demonstrated impressive performance.

Recently, interest has grown in the development of expert systems that reason "from first principles", i.e. from an understanding of the structure and function of the devices they are examining. This approach has been explored in a number of domains, including medicine (Patil, Szolovits & Schwartz, 1981), computer-aided instruction (Brown, Burton & deKleer, 1982), and electronic troubleshooting (Davis *et al.*, 1982; Genesereth, 1981), with the "devices" ranging from the gastro-intestinal tract, to transistors

1

DEVELOPMENTS IN EXPERT SYSTEMS
ISBN 0-12-187580-6

and digital logic components like adders or multiplexors. Our work has focused on the last of these domains, attempting to build a troubleshooter for digital electronic hardware.

In keeping with the notion of reasoning from first principles, we seek to build a system capable of reasoning in a fashion similar to an experienced electrical engineer. In particular, we wish to capture the skill exhibited by an engineer who can troubleshoot a device by reference to its schematics even though he may never have seen that particular device before. To do this we require something more than a collection of associations specific to a given device. We will see that the alternative mechanism is both device independent and revealing for what it indicates about the nature of the diagnostic process.

We view the task as a process of reasoning from behavior to structure, or more precisely, from misbehavior to structural defect. We are typically presented with a machine exhibiting some form of incorrect behavior and must infer the structural aberration that is producing this. The task is interesting and difficult because the devices we want to examine are complex and because there is no well developed theory of diagnosis for them.

The initial focus of our work has been to develop three elements that appear to be fundamental to our overall task. We require (i) a language for describing structure, (ii) a language for describing function, and (iii) a set of principles for troubleshooting that uses the two descriptions to guide its investigation.

This paper describes each of those components, paying particular attention to the nature of the reasoning that underlies the diagnostic process. We describe how that reasoning works and how it differs fundamentally from an approach built on empirical associations.

Section 2 provides an overview of our approach to describing structure and behavior. Sections 3–6 deal with troubleshooting. In section 3 we show why the traditional approach to reasoning about circuits—test generation—solves a different problem and we discuss a number of its practical shortcomings. We consider next the style of debugging known as discrepancy detection and demonstrate why it is a fundamental advance over traditional test generation. Further exploration, however, demonstrates that discrepancy detection in its standard form offers relatively little help in dealing with some common classes of faults. We suggest that the difficulty arises from a number of interesting assumptions made implicitly when using the technique.

In discussing how to repair the problems uncovered (sections 4 and 5), we argue for the primacy of *models of causal interaction*, rather than the traditional fault models. We point out the importance of making these models explicit, separated from the troubleshooting mechanism, and retractable in much the same sense that inferences are retracted in current systems.

Section 6 works through the diagnosis of a bridge fault using the machinery we have developed. We demonstrate how the careful organization of the interaction models allows us to make simplifying assumptions initially, surrendering them gracefully to consider more complex hypotheses when necessary.

In section 7 we describe the difficulties that would be encountered in attempting to use a traditional rule-based system to solve the same problem. We then explore the appropriate role of empirical associations in a system oriented toward reasoning from first principles.

2. Describing structure and function

If we wish to reason from knowledge of structure and function, we need a way of describing both. We have developed representations for each of these, described in detail elsewhere (Davis *et al.*, 1982; Davis & Shrobe, 1983). We limit our description here to reviewing only those characteristics of our representations important for understanding what follows.

2.1. FUNCTIONAL ORGANIZATION, PHYSICAL ORGANIZATION

By structure we mean information about the interconnection of modules. Roughly speaking, it is the information that would remain after removing all the textual annotation from a schematic.

Two different ways of organizing this information are particularly relevant to machine diagnosis: functional and physical. The functional view gives us the machine organized according to how the modules interact; the physical view tells us how it is packaged. We thus prefer to replace the somewhat vague term "structure" by the more precise terms *functional organization* and *physical organization*. In our system every device is described from both perspectives, producing two distinct (but interconnected) descriptions.

The basic unit of description is a *module*, similar in spirit to the notion of a black box (Fig. 1). Modules have *ports*, the places through which information enters and leaves the module. Every port has at least two terminals, one terminal on the outside of the port and one or more inside. Terminals are primitive elements; they store data representing the information flowing into or out of a device through their port, but are otherwise devoid of substructure.

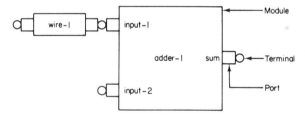

FIG. 1. The basic terms used in structure description.

Two modules are attached to one another by superimposing their terminals. In Fig. 1, for example, wire-1 is a (wire) module that has been attached to input-1 of adder-1 (an adder module) in this fashion.

Both the functional and physical descriptions are hierarchical in the usual sense: modules at any level may have substructure. An adder, for example, can be described by a functional hierarchy (adder, individual bit slices, half-adders, primitive gates) and a physical hierarchy (cabinet, board, chip). The two hierarchies are interconnected, since every primitive module appears in both: a single xor-gate for example, might be both functionally part of a half-adder, which is functionally part of a single bitslice of an adder, etc., and physically part of chip E67, which is physically part of board 5, etc. Cross-link information for primitive modules is supplied by the schematic; additional cross-links can be inferred by intersection (e.g. the adder can be said to be on board 3 because all of its primitive components are in chips on board 3).

Figure 2 shows the next level of structure of the adder and illustrates why ports may have multiple terminals on their inside: ports provide the important function of shifting level of abstraction. It may be useful to think of the information flowing along wire-1 as an integer between 0 and 15, yet we need to be able to map those four bits into the four single-bit lines inside the adder. Ports are the place where such information is kept.

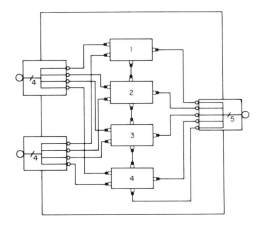

FIG. 2. Next level of structure of the adder.

As described in Davis & Shrobe (1983), the structural description of a module is expressed as a set of commands for building the module. These commands are executed by the system, creating data structures that model the components and their interconnections shown.

Our description language has been built on a foundation provided by a subset of DPL (Batali & Hartheimer, 1980). While DPL as originally implemented was specific to VLSI design, it proved relatively easy to "peel off" the top level of language (which dealt with chip layout) and rebuild on that base the new layers of language described above.

Since pictures are a fast, easy and natural way to describe structure, we have developed a simple circuit drawing system that permits interactive entry of pictures like those in Figs 1 and 2. Circuits are entered with a combination of mouse movements and key strokes; the resulting structures are then "parsed" into the language described Davis & Shrobe (1983).

2.2. BEHAVIOR DESCRIPTION

A variety of techniques have been explored in describing behavior, including simple rules for mapping inputs to outputs, petri nets, and unrestricted chunks of code. Simple rules are useful where device behavior is uncomplicated, petri nets are useful where the focus is on modeling parallel events, and unrestricted code is often the last resort when more structured forms of expression prove too limited or awkward. Various combinations of these three have also been explored.

Our initial implementation is based on a constraint-like approach (Sussman & Steele, 1980). Conceptually a constraint is simply a relationship. The behavior of the adder of Fig. 1, for example, can be expressed by saying that the logic levels of the terminals on ports *input-1, input-2* and *sum* are related in the obvious fashion.

In practice, this is accomplished by defining a set of rules covering all different computations (the three for the adder are shown below) and setting them up as demons that watch the appropriate terminals. A complete description of a module, then, is composed of its structural description as outlined earlier and a behavior description in the form of rules that interrelate the logic levels at its terminals.

> to get sum from (input-1 input-2) do (+ input-1 input-2)
> to get input-1 from (sum input-2) do (− sum input-2)
> to get input-2 from (sum input-1) do (− sum input-1)

A set of rules like this is in keeping with the original conception of constraints, which emphasized the non-directional, relationship character of the information. When we attempt to use it to model causality and function, however, we have to be careful. This approach is well suited to modeling causality and behavior of analog circuits, where devices are largely non-directional. But we can hardly say that the last two rules above are a good description of the *behavior* of an adder chip—the device doesn't do subtraction; putting logic levels at its output and one input does not cause a logic level to appear on its other input.

The last two rules really model the *inferences we make about the device*. Hence we find it useful to distinguish between rules representing *flow of electricity* (digital behavior, the first rule above) and rules representing *flow of inference* (conclusions we can make about the device, the next two rules). This not only keeps the representation "clean", but as we will see, it provides part of the foundation for the troubleshooting mechanism.

As described in more detail in Davis & Shrobe (1983), we have elaborated the basic mechanism above to deal with devices with memory. While our model of time is currently very simple, it allows us to describe the behavior of flip-flops, shift registers, and memory cells.

A set of constraints is a relatively simple mechanism for specifying behavior, in that it offers no obvious support for expressing behavior that falls outside the "relation between terminals" view. The approach also has known limits. Propagating values, for example, works well when dealing with simple quantities like numbers or logic levels, but runs into difficulties if it becomes necessary to work with symbolic expressions.†
The approach has, nevertheless, provided a good starting point for our work, in particular, permitting us to build the troubleshooting system described in section 3.

---

† What, for example, do we do if we know that the output of an or-gate is 1 but we don't know the value at either input? We can refrain from making any conclusion about the inputs, which makes the rules easy to write but misses some information. Or we can write a rule which express the value on one input in terms of the value on the other input. This captures the information but produces problems when trying to use the resulting symbolic expression elsewhere. See Kelley & Steinberg (1982) for an example of one plausible approach to this problem.

3. Troubleshooting

Having provided a way of describing functional organization, physical organization and behavior, we come now to the important third step of providing a diagnostic mechanism that works from those descriptions. We develop the topic of troubleshooting in three stages. We begin by considering test generation, the traditional approach to automated reasoning about circuits, and explain how it falls short of our requirements.

We consider next the style of debugging we call *discrepancy detection* and demonstrate why it is a fundamental advance. Further exploration, however, demonstrates that in its standard form this approach encounters interesting limits in dealing with commonly known classes of faults. We suggest that the problem arises from a number of implicit assumptions typically made when using the technique.

In discussing how to repair the problems uncovered, we argue for the primacy of *models of causal interaction*, rather than traditional fault models. We point out the importance of making these models explicit, separate from the troubleshooting mechanism, and retractable.

This leads us to a demonstration of the power of this approach on a bridge fault, a traditionally difficult problem. We show how the careful and explicit layering of models of interaction makes it possible to diagnose and locate the fault in a sharply focused process that generates only a few plausible candidates.

3.1. THE TRADITIONAL APPROACH

The traditional approach to troubleshooting digital circuitry [for example, Breuer & Friedman (1976)] relies primarily on the process of path sensitization in a range of forms, of which the D-algorithm (Roth, 1966) is one of the most powerful. A very simple example of path sensitization will illustrate the essential character of the process.

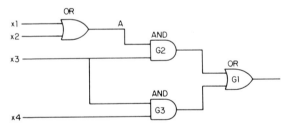

FIG. 3. Simple example of path sensitization.

Consider the circuit shown in Fig. 3 and imagine that we want to determine whether the wire labeled A is stuck at 1. We try to put a zero on it by setting both x1 and x2 to 0. Then, to observe the actual value on the wire we set x3 to 1, thereby propagating the value unchanged through G2, and set x4 to 0, making the output of G3 = 0, allowing the value of A to propagate unchanged through G1.

For our purposes this approach has a number of significant drawbacks. Perhaps most important, it is a theory of *test generation*, not a theory of *diagnosis*. Given a specified fault, it is capable of determining a set of input values that will detect the fault (i.e. a set of values for which the output of the faulted circuit differs from the output of a good circuit). The theory tells us how to move from faults to sets of inputs; it provides little help in determining what fault to consider, or which component to suspect.

These questions are a central issue in our work for several reasons. First, the level of complexity we want to deal with precludes the use of diagnosis trees—a complete decision tree for all possible faults—since they quickly become computationally infeasible. Second, our basic task is repair, rather than initial testing. Hence the problem confronting us is "Given the following piece of misbehavior, determine the fault". We are not asking whether a machine is free of faults, we know that it fails and know how it fails. Given the complexity of the device, it is important to be able to use this information as a focus for further exploration.

A second drawback of the existing theory is its use of a set of explicitly enumerated faults. Since the theory is based on boolean logic, it is strongly oriented toward faults whose behavior can be modeled as some form of permanent binary value, typically the result of stuck-ats and opens. One consequence of this is the paucity of useful results concerning bridging faults.

3.2. DISCREPANCY DETECTION

One response to these problems has been the use of what we may call the "discrepancy detection" approach (for example, deKleer, 1976; Brown *et al.*, 1982). The basic insight of the technique is the substitution of violated expectations for specific fault models. That is, instead of postulating a possible fault and exploring its consequences, the technique simply looks for mismatches between the values it expected from correct operation and those actually obtained. This allows detection of a wide range of faults because misbehavior is now simply defined as anything that isn's correct, rather than only those things produced by a stuck-at on a line.

This approach has a number of advantages. It is, first of all, fundamentally a diagnostic technique, since it allows systematic isolation of the possibly faulty devices, and does so without having to precompute fault dictionaries, diagnosis trees, or the like. Second, it appears to make it unnecessary to specify a set of expected faults (we comment further on this below). As a result, it can detect a much wider range of faults, including any systematic misbehavior exhibited by a single component. The approach also allows natural use of hierarchical descriptions, a marked advantage for dealing with complex structures.

This approach is a good starting point, but has a number of important limitations built into it. We work through a simple example to show the basic idea and use the same example to comment on its shortcomings.

Consider the circuit in Fig. 4.† If we set the inputs as shown, the behavior descriptions will indicate that we should expect 12 at F. If, upon measuring, we find the value at F to be 10, we have a conflict between observed results and our model of correct behavior. We employ the notion of dependency-directed backtracking (Stallman & Sussman, 1977) to enumerate the possible sources of the problem. We check the dependency record at F to find that the value expected there was determined using the behavior rule for the adder and the values emerging from the first and second multiplier. One of those three must be the source of the conflict, so we have three

---

† As is common in the field, we make the usual assumptions that there is only a single source of error and the error is not transient. Both of these are important in the reasoning that follows.

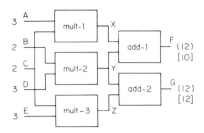

FIG. 4. Troubleshooting example using discrepancy detection. ( ) Expected, [ ] actual result.

possibilities to pursue: either the adder behavior rule is inappropriate (i.e. the first adder is broken), or one of the two inputs did not have the expected values (and the problem lies further back).

Consideration of the first possibility immediatly generates hypothesis #1: adder-1 is broken. Note also that this process yields symptom information about the specific malfunction. If adder-1 is indeed the culprit, it is receiving a 6 on each of its inputs but producing a 10 at its output. We thus know a little bit about how the module is failing; the utility of this information is illustrated in section 6.

To pursue the second possibility, we assume that the second input to adder-1 is good. In that case the first input must have been a 4 (reasoning from the result at F, valid behavior of the adder, and one of the inputs), but we expected a 6. Hence we now have a discrepancy at the input to adder-1; we have succeeded in pushing the discrepancy one step further back. The expected value there was based on the behavior rule for the multipler and the expected value of its inputs. Since the inputs to the multiplier are primitive (supplied by the user), the only alternative along this line of reasoning is that the multiplier is broken. Hence hypothesis #2 is that adder-1 is good and multiplier-1 is faulty.

Pursuing the third possibility: if the first input to adder-1 is good, then the second input must have been a 4 (suggesting that the second multiplier might be bad). But if that were a 4, then the expected value at G would be 10 (reasoning forward through the second adder). We can check this and discover in this case that the output at G is 12. Hence the value on the output of the second multiplier can't be 4, it must be 6, so the second multiplier can't be causing the current problem.

This approach is a useful beginning. Note, for example, that it is diagnostic: it enumerates the devices that could have caused the symptoms noted. It also reasons from the structure and function of the device: the candidate generation process works from the schematic itself to determine which components might be to blame.

3.3. WHERE DISCREPANCY DETECTION FALLS SHORT

But the approach also has some clear shortcomings. Consider the slightly revised example shown in Fig. 5. Reasoning as before, we would discover in this case that there is only one hypothesis consistent with the values measured at F and G: the second multiplier is malfunctioning, outputting a 0.

Yet there is another quite reasonable hypothesis: the third multiplier might be bad (or the first).

But how could this produce errors at both F and G? The key lies in being wary of our models. The thought that digital devices have input and output ports is a convenient

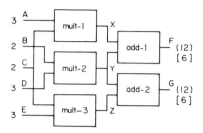

FIG. 5. Troublesome troubleshooting example. ( ) Expected, [ ] actual results.

abstraction, not an electrical reality. If, as sometimes happens (due to a bent pin, bad socket, etc.), a chip fails to get power, its inputs are no longer guaranteed to act unidirectionally as inputs. If the third multiplier were a chip that failed to get power, it might not only send out a 0 along wire z, but it might also pull down wire C to 0. Hence the symptoms result from a single point of failure (multiplier-3), but the error propagates along an "input" line common to two devices.

The immediate problem with this straightforward use of discrepancy detection lies in its implicit acceptance of unidirectional ports and the reflection of that acceptance in the basic dependency-unwinding machinery. We implicitly assumed that wires get information only from output ports—when checking the inputs to multiplier-1, we assume that the inputs are "primitive". We looked only at terminals A and C, never at the other end of the wire at multiplier-3.

Bridges are a second common fault that illustrates an interesting shortcoming in the approach: the reasoning style used above can *never* hypothesize a bridging fault, again because of implicit assumptions about the model and their subtle reflection in the method. Bridges can be viewed as wires that don't show up in the design. But the traditional approach makes an implicit closed world assumption—the structure description is assumed to be complete and anything not shown there "doesn't exist". Clearly this is not always true. Bridges are only one manifestation; wiring errors during assembly are another possibility.

Let's review for a moment. The traditional test generation technology suffered from two problems: it is a technology for test generation, not diagnosis, and it uses a very limited fault model. The discrepancy detection approach improves on this substantially by providing a diagnostic ability and by defining a fault as anything that produces behavior different from that expected. This seems to be perfectly general, but, as we illustrated, it is in fact limited in some important ways. We believe it is instructive to examine the basic source and nature of those limitations.

4. Models of interaction

One common claim about the discrepancy detection approach is that it makes no assumptions about the character of the fault. Yet our counterexamples show this not to be the case. What is the source of the problem? We believe that the issue is not in fact the character of the fault models, it is assumptions about *models of causal interactions.* In the port problem, for example, the assumption was that there was only one possible *direction of causality* at an input port. For the bridge problem, the assumption was that the only possible *paths of interaction* were the wires shown in the diagram. These are assumptions about possible pathways of causality, not categories of faults.

The problem is not in the existence of such assumptions—they are, in fact, crucial to the diagnostic process. The problem lies instead in the careful and explicit management of them. To see the necessity of having assumptions about causal pathways, consider the nature of the candidate generation task. Given a problem noticed at some point in the device, candidate generation attempts to determine which modules could have caused the problem. To answer the question we must know by what mechanisms and pathways modules can interact. Without *some* notion of how modules can affect one another, we can make no choice, we have no basis for selecting any one module over another.

The obvious answer in this domain is "wires": modules interact because they're explicitly wired together. But that's not the only possibility. Bridges, as we saw, are an exception; they are "wires" that aren't supposed to be there. But we also might consider thermal interactions, capacitive coupling, transmission line effects, and other possible pathways.

Our task then, in generating possible candidates, is not to trace wires, it is to *trace paths of causality*. Wires are only the most obvious pathway, they are by no means the only one. In fact, given the wide variety of faults we'd like to be able to deal with, we need many different models of interaction.

And that leaves us on the horns of a classic dilemma. If we omit any interaction model, there will be whole classes of faults we will never be able to diagnose. Yet if we include every model, our candidate generation becomes virtually indiscriminate—there will be some (possibly convoluted) pathway by which every module could conceivably be to blame.

What can we do? We believe that two steps are important. First, we have to recognize that our inference machinery—in this case discrepancy detection and dependency-directed backtracking—is not the source of problem-solving power. Discrepancy detection is, as we have noted, an important advance because it is diagnostic and because it allows use of the description hierarchy. But the mechanism is only as powerful as the models we supply. And therein lies the second step. The difficult and important work is the enumeration and careful organization of the models of interaction.

5. Layering interaction models

We believe there is a way out of the problem based on a careful enumeration and layering of the interaction models. The basic idea is quite simple. List all of the models and order them by some appropriate metric (more on this below). Start the diagnosis using only the first model and add additional ones only if we reach an intractable contradiction (i.e. given the current model(s) and set of assumptions, there is no way to account for the observed behavior).

A plausible guess at an ordering for the models might be

localized failure of function (e.g. stuck-at on a wire, failure of a RAM cell),
bridges,
unexpected direction (e.g. the power failure problem),

multiple point of failure,
assembly error,
design error.

We would start with the assumptions that the structure is as shown in the schematic, that there was only a single point of error, that information flowed only in the predicted directions, etc., and attempt to generate candidates in the localized failure category. Only if this led to a contradiction would we be willing to surrender an assumption (e.g. that the schematic was correct) and entertain the notion that a bridge might be at fault. If this, too, led us down a blind alley then we would surrender additional assumptions and consider ever more elaborate hypotheses, eventually entertaining the possibility of multiple errors, an assembly error (every individual component works but they have been wired up incorrectly) and even design errors (the implementation is correct but cannot produce the desired behavior).

This mimics what we believe a good engineer will do: make all the assumptions necessary to simplify a problem and make it tractable, but be prepared to discover that some of those simplifications were incorrect. In that case, surrender some of those assumptions and be willing to consider additional models of failure, but retain any information gained in the previous step if it did not logically depend on the assumptions surrendered.

In terms of the dilemma noted above, the models serve as a set of filters. They restrict the categories of paths of interaction we are willing to consider, thereby preventing the candidate generation from becoming indiscriminate. But they are filters that we have carefully ordered and consciously put in place. If we cannot account for the observed behavior with the current filter in place, we remove it and replace it with one that is less restrictive, thereby allowing us to consider an additional category of interaction paths.

Clearly an important consideration in all this is the ordering metric. Currently we order the models by relying on the experience of expert troubleshooters, who tell us that some models of interaction are encountered more frequently than others. Stuck-ats are more likely than assembly errors, for example. While the ordering criteria may eventually need be more elaborate,† their precise content is less an issue here than its character: it is a summary of empirical experience that helps us to order the use of the interaction models. Were we willing to model enormously more of the world, we might be able to *infer* that design errors are less common in well-established designs, but this is beyond the scope of our efforts and the investment would be very large for a relatively modest return. As it stands, we are in this specific case willing to rely on some simple experiential observations for guidance, without bothering to model the causality in any detail.

6. Layers of interaction example: diagnosing a bridge fault

As we have noted, traditional automated reasoning about circuits works from a predefined list of fault models and uses the mathematical style of analysis exemplified

---

† We might want to indicate that design errors, for example, are less likely if the device has been around for a long time, or that assembly errors are more common in machines that arrived from the production line recently.

by boolean algebra or the D-algorithm. As a result, it is strongly oriented toward faults modelable as a permanent binary value. One problem with this is its inability to provide useful results concerning bridge faults.

In this section we show how our system diagnoses a bridge fault, illustrating a number of the ideas described above. The key point is the utility of layering the interaction models. We start with a very restricted model (failure of function) but soon discover that that leads us to a contradiction: there is no way to account for the observed behavior under that model. We surrender an assumption and consider an additional pathway of interaction: bridge faults. We show how knowledge of both structural and functional organization allows us to generate a select few bridge fault hypotheses, eventually discovering the underlying fault.

While the example has been simplified for presentation, there is still unfortunately a fair amount of detail necessary. A summary of the basic steps in our solution will help make clear how the problem is solved.

The device is a 6-bit carry-chain adder (Fig. 6). The problem begins when we notice that the attempt to add 21 and 19 produces an incorrect result.

The process outlined above generates a set S1 of three candidates whose malfunction can explain this result.

A new set of inputs (1 and 19) is chosen in an attempt to discriminate among the three possibilities. The adder's output is incorrect for this set of inputs also. The candidate generator indicates that there are two candidates capable of explaining this new result.

Neither of these two candidates are found in S1. Thus we reach a contradiction: no one component is capable of explaining the data from both sets of inputs.

Put slightly differently, we have a contradiction *under the current set of assumptions and interaction models*. We therefore have to surrender one of our assumptions and use a different interaction model.

The next model—bridge faults—surrenders the assumption that the structure is as shown in the schematic and considers one class of modifications to the structure: the addition of one wire between physically adjacent pins.

The combination of functional information (the expected pattern of values produced by the fault) and physical adjacency provides a strong constraint on the set of connections which might be plausible bridges.

The first application of this idea produces two hypotheses that are functionally plausible, but both are ruled out on physical grounds.

Dropping down a level of detail in our description reveals additional bridge candidates, two of which prove to be both physically and functionally plausible. One of these proves to be the actual error.

6.1. THE EXAMPLE

Consider the 6-bit carry-chain adder shown in Fig. 6 and imagine that the attempt to add 21 and 19 produces 36 rather than the expected value of 40. Invoking the candidate generation process described above, we find that there are three devices (SLICE-1, A2 and SLICE-2, highlighted in Fig. 6) whose malfunction can account for the misbehavior.†

---

† The example has been simplified slightly for the sake of presentation.

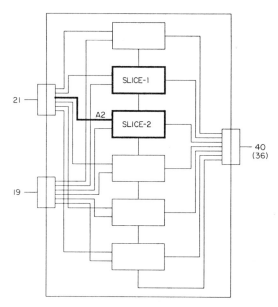

FIG. 6. Possibly faulty components as indicated by first test (heavy lines).

A good strategy when faced with several candidates is to devise a test that can cut the space of possibilities in half. In this case changing the first input (21) to 1 will be informative: if the output of SLICE-2 does not change (to a 0) when we add 1 and 19, then the error must be in either A2 or SLICE-2.[†]

As it turns out, the result of adding 1 and 19 is 4 rather than 20. Since the output of SLICE-2 has not changed, it appears that the error must be in either A2 or SLICE-2.

But if we invoke the standard candidate generation process, we discover an oddity: the only way to account for the behavior in which adding 1 and 19 produces a 4 is if one of the two candidates highlighted in Fig. 7 (B4 or SLICE-4) is at fault.

And therein lies an interesting contradiction. The only possible candidates that account for the behavior of the first test are the three in Fig. 6; the only possible candidates that account for the second test are the two in Fig. 7. There is no overlap, so there is no single candidate that accounts for all the observed behavior.[‡]

[†] This and subsequent test generation is currently done by hand; we are in the process of adding this capability to the system. The reasoning behind this test relies on the single fault assumption: if the malfunctioning component really were SLICE-1, both A2 and SLICE-2 would be fault-free. Hence the output of SLICE-2 would have to change when we changed one of its inputs. (Notice, however, if the output actually does change, we don't have any clear indication about the error location: SLICE-2, for example, might still be faulty.)

[‡] Note that there is nothing to be gained by dropping down another level of detail in the functional description. That cannot help resolve the contradiction because our functional description is a tree rather than a graph: in our work to date, at least, no component is used in more than one way, so the same component never shows up in more than one place in the functional hierarchy. (This is easily changed if it proves necessary: if the functional hierarchy were a graph, we would simply continue on down it to see if the two candidate sets did in fact have a subcomponent in common.) Hence if the two sets of candidates have nothing in common at one level of detail, they cannot have a subcomponent in common at a lower level.

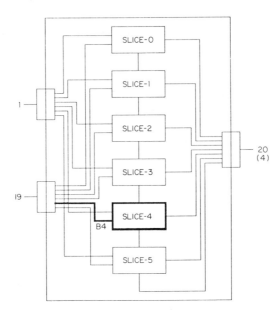

FIG. 7. Possibly faulty components as indicated by the second test.

Our current model—the localized failure of function—has thus led us to a contradiction. We therefore surrender our assumption that the structure is as shown in the schematic and consider an additional kind of interaction path—bridging faults. The problem now is to see if there is some way to unify the test results, some way to generate a single bridge fault candidate that accounts for all the observations.

It is useful to consider the nature of the problem we now encounter. As suggested earlier, bridges are in effect wires that are not supposed to be there. Much of the difficulty in dealing with them arises because they violate the important assumption that the structure of the device is as we expected it to be. Surrendering this assumption is only the beginning, however: we have only admitted we know what the structure *isn't.* Saying that we may have a bridge fault narrows it to a particular class of modifications we are willing to consider, but the real problem here remains one of *making plausible conjectures about modifications to the structure.* Between which two points can we insert a wire and produce the behavior observed?

To understand the nature of the machinery we use to answer that question, consider what we have and what we need. We have test results, i.e. behavior, and we want conjectures about modification to structure. The link from behavior to structure is provided by knowledge of electronics: in TTL, a bridge fault acts like an and-gate, with ground dominating.[†]

To see how to use this knowledge, consider the simple example shown in Fig. 8. Imagine that in Test 1, module MOD-1 should have produced a 1, but produced a 0

[†] This is, in fact, an oversimplification, but accurate enough to be useful. In any case, the point here is how the information is used; a more complex model could be substituted and carried through the rest of the problem.

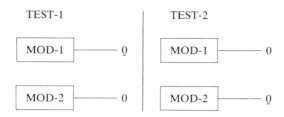

FIG. 8. Pattern of values indicative of a bridge.

(the zero is underlined to show that it is an incorrect output); MOD-2 should have produced a zero and did. This situation is exactly reversed in Test 2. The pattern displayed in these two tests makes it plausible that there is a bridge linking the outputs of MOD-1 and MOD-2.† In the first test the output of MOD-1 was dragged low by MOD-2, in the second test the output of MOD-2 was dragged low by MOD-1.

We have thus turned the insight from electronics into the simple pattern indicated above. It is plausible to hypothesize a bridge fault between two modules MOD-1 and MOD-2 if: in Test 1 MOD-1 produced an erroneous 0 and MOD-2 produced a valid 0, while in Test 2 MOD-1 produced a valid 0 while MOD-2 produced a erroneous 0.

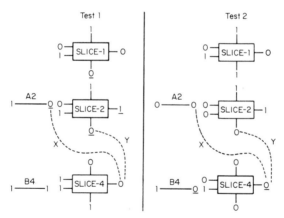

FIG. 9. Candidates and values at their ports.

Figure 9 shows the candidate generation results in somewhat more detail. As indicated in section 3, the candidate generation procedure can also indicate for each candidate the values that would have to exist at its ports for that candidate to be the broken one. For example, for SLICE-1 to be at fault in test 1, it would have to have the three inputs shown, with its sum output a zero (as expected) and its carry output also a zero (the manifestation of the error, underlined).

† In fact, if there is no such pattern anywhere in the observed behavior, then the problem cannot be a bridge fault.

The only way to account for all the test data shown in Fig. 9 with a single bridge fault is to have a bridge that connects one of the candidates generated by the first test (SLICE-1, A2, or SLICE-2) with one of the candidates generated by the second test (B4, SLICE-4). Thus our problem becomes one of finding the pattern described above with MOD-1 in the first set of candidates and MOD-2 in the second set.

In Fig. 9, there are two such pairs:

broken line X, bridging wire A2 to the sum output of SLICE-4; and
broken line Y, bridging the carry-out of SLICE-2 to the sum output of SLICE-4.

The values on those devices match the desired pattern, so either one is a functionally plausible bridge.

But, as is well known, there is an important additional constraint on bridging faults: they have to be physically plausible as well. For the sake of simplicity, let us assume here that bridge faults result only from solder splashes at the pins of chips.† To check the physical plausibility of our hypothesis, we switch to our physical representation

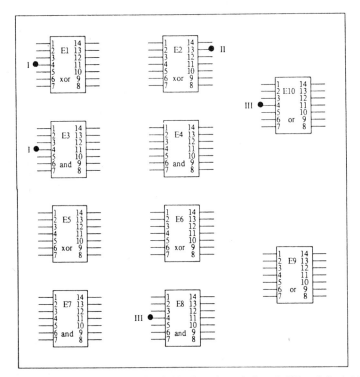

FIG. 10. Physical layout of the board with first bridge hypotheses indicated. (Slices 0, 2, and 4 are in the upper five chips, slices 1, 3, and 5 are in the lower five.) I, End of wire A2; II, sum output of SLICE-4; III, carry-out of SLICE-2.

† Again this is correct but oversimplified: backplane pins can be bent or bridged by dropped pieces of metal, backplane wiring insulation can be cut by too tight a turn around backplane pins, etc. As before, we can introduce this more complex model if necessary without change to the procedure above.

(Fig. 10). Wire A2 is connected to chip E1 at pin 4 and chip E3 at pin 4; the sum output of SLICE-4 emerges at chip E2, pin 13. Since they are not adjacent, the first hypothesis is not physically reasonable. Similar reasoning rules out Y, the hypothesized bridge between the carry-out of SLICE-2 and the sum output of SLICE-4.

But thus far we have only the considered top level of functional organization of the adder. If we take each of the components shown in Fig. 9 and drop down one level of structural detail,† we obtain the components and values shown in Fig. 11. Checking here for the desired pattern, we find that either of the two points labeled A2 and S2 could be bridged to either of the two points labeled S4 and C4, generating four functionally plausible bridge faults.

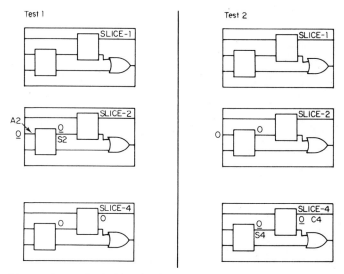

Fig. 11. Candidates at the next level of functional description. To simplify the drawing, we show only the relevant values.

Once again we check physical plausibility by examining the actual location of A2, S2, S4, and C4 (Fig. 12). As illustrated there, two bridge faults are physically possible: A2–S4 and S2–S4.

Switching back to our functional organization once more (Fig. 13), we see that the two possibilities correspond to (X) an output-to-input bridge between the xor gates in the rear half-adders of SLICE-2 and SLICE-4, and (Y) a bridge between two inputs of the xor in the forward half-adders of SLICES-2 and 4.

We can distinguish between these two possibilities by adding 0 and 4.‡ With these as primary inputs, the inputs of SLICE-2 will be 1 and 0, with a carry-in of 0, while the inputs of SLICE-4 will both be 0, with a carry-in of 0. This set of values will show the effects of bridge Y, if it in fact exists: the sum output of SLICE-2 will be a 0 if it

† Dropping down a level of detail proves useful here because additional substructure becomes visible, effectively revealing new places that might be bridged.

‡ As before, the test generation is currently done by hand.

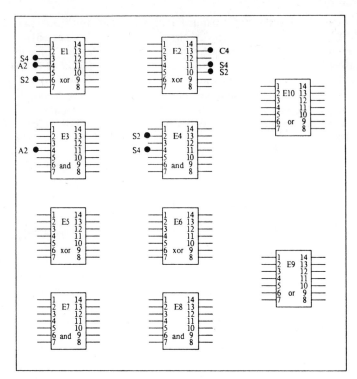

FIG. 12. Second set of bridge hypotheses located on physical layout.

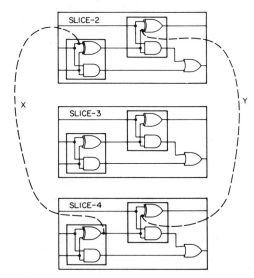

FIG. 13. Functional representation with bridge fault hypotheses illustrated.

does exist and a 1 otherwise. When we perform this test the result is a 1, hence bridge Y is not in fact the problem.

Bridge X becomes the likely answer, but we should still test for it directly. Adding 4 and 0 (i.e. just switching inputs A and B), is informative: if bridge X exists the result will be 0 and a 1 otherwise. In this case the result is 0, hence the bridge labeled X is in fact the problem.†

7. Why not a traditional rule-based system?

Given our emphasis in this paper on the use of structural and causal models as underpinnings for troubleshooting, it is worth asking whether a more familiar approach to expert system diagnosis—a rule-based system—would do as well in diagnosing computer malfunction. We believe there are considerable difficulties in using such an approach.

It is important first to ask what we mean when we suggest using a traditional rule-based system. The question is not simply whether there are conditional statements anywhere in the program; the central issue is the character of the knowledge encoded by those statements. As we noted earlier, rule-based systems have traditionally been constructed from collections of empirical associations that link symptoms and diseases. The question is really, how does a system built from rules *of that sort* compare to the present effort?

The primary shortcomings of such a system are the difficulty of initial construction and the lack of carry-over to the later efforts. The system is difficult to construct because the rules must in general be extracted on a case by case basis. If the symptoms are external manifestations of misbehavior (e.g. machine halted, parity error, disk errors, etc.), the best we can do is to ask an expert for the rules dealing with each case. No more systematic method of collecting rules is available and the process typically continues for an extended period of time.

The extensive effort might still be acceptable if it could be used as a foundation for later systems, but the second problem precludes that: there is distressingly little carry-over from one machine to the next when we capture troubleshooting knowledge at the level of symptom/disease associations. Even simple changes to a single machine (e.g. design upgrades) can mean substantial changes in the rules. Working on a new machine means even more difficulty: the rules for a DEC 2020 would be very different from those for a 2060, for example. Each new system would require a new knowledge base with all of the difficulties that entails.

Yet there is much in common across that range of machines: concepts like the organization of computers (memory, buses, etc.) and the basics of digital logic would surely be useful in understanding and debugging all of them. Even more fundamentally, concepts like modules linked by causal pathways, expected behavior, etc., are valid across the range of devices mentioned and extend considerably further, to include handheld calculators, digital watches, automobiles and software.

There is thus substantially greater opportunity for carry-over in an approach that bases the troubleshooting process on an understanding of "how the device works". It

---

† Had both been ruled out by direct test, then we would once again have had a contradiction on our hands and would have had to drop back to consider yet a more elaborate model with additional paths of interaction.

can also be substantially easier to construct, since we do have a way of systematically enumerating the required knowledge: the structure and function embodied in the design of the device. Modifications to the design are also more easily accommodated: we can update the structure and function specifications directly, rather than having to determine their consequences on behavior and troubleshooting.

The issue then is the content and character of the knowledge. We find it considerably more useful to organize the knowledge around and work from descriptions of structure and function, rather than associations linking symptoms and diseases.

8. Summary

We have briefly described our work to date on developing languages and mechanisms for describing structure and function in digital circuitry. The techniques we have developed offer a number of useful advantages, including distinguishing clearly between design and implementation.

We traced briefly the evolution of atuomated troubleshooting. We noted that the traditional technology focuses on test generation, which is only a small part of the diagnostic task. We saw that the discrepancy detection approach offers a significant advance, but provides little assistance for a number of important classes of faults. We found that those limits arise from some subtle and important assumptions about the nature of the causal interactions between modules.

The basic idea behind discrepancy detection is keeping careful records of the reasons for our expectations and then backtracking through those reasons to uncover the source of any discrepancy between expectations and observations. This gives it its ability to generate candidates, but is also the source of its limitations. If one of the reasons for an expectation is "the structure of the device is as shown", then admitting that that assumption might be false is only the first step. Proposing a plausible modification is the difficult part. The discrepancy detection technique leads us to consider the need to modify our structure description, but it tells us little about how to proceed from there.

Our response was to assert the primacy of models of causality rather than models of faults and to focus on the enumeration and organization of interaction models. We are building a system in which the complexity of the problem is handled by invoking the models explicitly, by layering them carefully, and by being able to retract one model and substitute another in the event of failure.

The utility of this approach was demonstrated in the diagnosis of a bridge fault in which relatively few, carefully chosen hypotheses were generated.

We have significant work yet to do in in a number of directions. Our behavior description language, for example, is still quite basic and needs considerable elaboration to deal with more complex devices. Attention also needs to be paid to determining a more complete list of models and to exploring the difficulties that will no doubt be encountered in dealing with more complex problems like multiple faults, assembly errors, etc. But we feel that the current approach has proven to be a sound and useful foundation for creating more interesting and powerful diagnostic reasoners.

Contributions to this work we made by all of the members of the Hardware Troubleshooting project at M.I.T., including: Howie Shrobe, Walter Hamscher, Karen Wieckert, Mark Shirley, Harold Haig, Art Mellor, John Pitrelli, and Steve Polit.

Michael Coombs and Patrick Winston provided comments on earlier drafts, contributing a number of useful suggestions on presentation.

This report describes research done at the Artificial Intelligence Laboratory of the Massachusetts Institute of Technology. Support for the laboratory's Artificial Intelligence research on electronic troubleshooting is provided in part by a grant from the Digital Equipment Corporation.

References

BATALI, J. & HARTHEIMER, A. (1980). The design procedure language manual. *Massachusetts Institute of Technology, Artificial Intelligence Memo 598.*

BREUER, M. & FRIEDMAN, A. (1976). *Diagnosis and Reliable Design of Digital Systems.* Rockville, Maryland: Computer Science Press.

BROWN, J. S., BURTON, R. & DEKLEER, J. (1982). Pedagogical and knowledge engineering techniques in SOPHIE I, II and III. In SLEEMAN, D. H. & BROWN, J. S., Eds, *Intelligent Tutoring Systems.* London: Academic Press.

DAVIS, R. (1982). Expert systems: where are we and where do we go from here? *The AI Magazine* (spring), 3–22.

DAVIS, R. & SHROBE, H. E. (1983). Representing structure and behavior of Digital Hardware, *IEEE Spectrum* (to appear).

DAVIS, R., AUSTIN, H., CARLBOM, I., FRAWLEY, B., PRUCHNIK, P., SNEIDERMAN, R. & GILREATH, J. (1981). The dipmeter advisor: interpretation of geologic signals. *Proceedings of the International Joint Conference on Artificial Intelligence*, pp. 846–849.

DAVIS, R., SHROBE, H., HAMSCHER, W., WIECKERT, K., SHIRLEY, M. & POLIT, S. (1982). Diagnosis based on description of structure and function. *Proceedings of the American Association for Artificial Intelligence*, pp. 137–142.

DEKLEER, J. (1976). Local methods for localizing faults in electronic circuits. *Massachusetts Institute of Technology, Artificial Intelligence Memo 394.*

DEKLEER, J. & BROWN, J. S. (1982). Assumptions and ambiguities in mechanistic mental models. *Xerox PARC Report CIS-9.*

GASCHNIG, J. (1979). Preliminary evaluation of the performance of the PROSPECTOR system for mineral exploration. *Proceedings of the International Joint Conference on Artificial Intelligence*, pp. 308–310.

GENESERETH, M. (1981). The use of hierarchical models in the automated diagnosis of computer systems. *Stanford HPP Memo 81-20.*

KELLEY, V. & STEINBERG, L. (1982). The CRITTER system—analyzing digital circuits by propagating behaviors and specifications. *Proceedings of the AAAI Conference*, pp. 284–289 (August).

PATIL, R., SZOLOVITS, P. & SCHWARTZ, W. (1981). Causal understanding of patient illness in medical diagnosis. *Proceedings of the International Joint Conference on Artificial Intelligence*, pp. 893–899.

POPLE, H. (1982). Heuristic methods for imposing structure on ill-structured problems. In SZOLOVITS, P., Ed., *Artificial Intelligence in Medicine. (AAAS Selected Symposium 51.)*

ROTH, J. P. (1966). Diagnosis of automata failures: a calculus and a method. *IBM Journal of Research and Development*, 278–291.

SHORTLIFFE, E. (1976). *Computer-Based Medical Consultations: MYCIN.* New York: American Elsevier.

STALLMAN, R. M. & SUSSMAN, G. J. (1977). Forward reasoning and dependency-directed backtracking in a system for computer-aided circuit analysis. *Artificial Intelligence*, **9**, 135–196.

SUSSMAN, G. & STEELE, G. (1980). Constraints—a language for expressing almost-hierarchical descriptions. *Artificial Intelligence*, **4**, pp. 1–40.

Deep versus compiled knowledge approaches to diagnostic problem-solving†

B. Chandrasekaran and Sanjay Mittal‡

Department of Computer and Information Science, The Ohio State University, Columbus, Ohio 43210, U.S.A.

Most of the current generation expert systems use knowledge which does not represent a deep understanding of the domain, but is instead a collection of "pattern → action" rules, which correspond to the problem-solving heuristics of the expert in the domain. There has thus been some debate in the field about the need for and role of "deep" knowledge in the design of expert systems. It is often argued that this underlying deep knowledge will enable an expert system to solve hard problems. In this paper we consider diagnostic expert systems and argue that given a body of underlying knowledge that is relevant to diagnostic reasoning in a medical domain, it is possible to create a diagnostic problem-solving structure which has all the aspects of the underlying knowledge needed for diagnostic reasoning "compiled" into it. It is argued this compiled structure can solve all the diagnostic problems in its scope efficiently, without any need to access the underlying structures. We illustrate such a diagnostic structure by reference to our medical system MDX. We also analyze the use of these knowledge structures in providing explanations of diagnostic reasoning.

1. Introduction

Recently Hart (1982) and Michie (1982) have written about the "depth" at which knowledge is represented and used in problem-solving by expert systems. Hart makes a distinction between "deep" and "surface" systems, which is similar to the distinction between "high road" vs "low road" approaches discussed by Michie. The underlying idea is that surface systems are at best a data base of pattern–decision pairs, with perhaps a simple control structure to navigate through the data base. MYCIN (Shortliffe, 1976) would be an example of such a system. There is less agreement on exactly what characterizes deep systems, but it is suggested that deep systems will solve problems of significantly greater complexity than surface systems can. This distinction appears to capture a fairly widespread feeling about the inadequacy of a variety of first generation expert systems.

In the area of medical diagnosis—which will be the exclusive concern of this paper, even though the spirit of what we say may be applicable to other tasks as well—the straightforward approach of building a data base of patterns relating data and diagnostic states is not feasible given the large number of patterns that would be needed in any realistic medical domain. There is also the pragmatic problem of coming up with a complete set of such patterns in the first place. The next best approach, namely, devising some problem-solving mechanism which operates on a data base of partial patterns (i.e. patterns relating only a small set of data to diagnostic states; also called

---

† This is an extended version of a paper of the same title which was presented at the 1982 Conference of American Association for Artificial Intelligence, Pittsburgh, Pennsylvania, U.S.A.

‡ *Current address:* Knowledge Systems Area, Xerox PARC, 3333 Coyote Hill Road, Palo Alto, California 94304, U.S.A.

DEVELOPMENTS IN EXPERT SYSTEMS
ISBN 0-12-187580-6

situation–action rules) has been tried with moderate success: e.g. MYCIN. The above two approaches fall into the category of surface systems.

The major intuition behind the feeling that expert systems should have deep models is the observation that often even human experts resort to "first principles" when confronted with an especially knotty problem. Also, there is the empirical observation that a human expert who cannot explain the basis for his reasoning by appropriate reference to the deeper principles of his field will have credibility problems, especially in life and death areas such as medicine. Added to this is the often unspoken assumption that the speed and efficiency with which an expert solves problems can be accounted for by hypothesizing that in most cases the physician uses a data base of commonly occurring patterns similar to that described earlier for rapid problem-solving. In this view a simple data base of patterns is no longer adequate in hard cases, and invocation of deeper structures is needed.

These considerations have resulted in calls for the representation and manipulation of deeper knowledge structures in expert system design. There is, however, no general agreement on the form and content of these deeper structures. Several alternatives are conceivable: mathematical and simulation models of a complex process, physical laws that govern a situation, functional models (deKleer & Brown, 1981) of how a device works, detailed causal networks, and collections of rules of the form "If in ⟨situation⟩, ⟨action⟩ is taken, ⟨situation⟩ will follow" (Michie, 1982). The last-mentioned proposal is based on the idea that understanding a domain often corresponds to an ability to deduce consequences of events that may occur in the domain. We discuss in a later section of the paper the relation between the depth of a domain model and ability to derive causal connections. For our immediate purposes, it is not necessary to make a commitment to one or another of these candidates, since our arguments will not make any specific assumptions about the nature of the deeper representation.

The thesis of this paper is as follows. Between the extremes of a data base of patterns on one hand and representations of deep knowledge (in whatever form) on the other, there exists a knowledge and problem-solving structure which (1) has all the relevant deep knowledge "compiled" into it in such a way that it can handle all the diagnostic problems that the deep knowledge is supposed to handle if it is explicitly represented and used in problem-solving; and (2) will solve the diagnostic problems more efficiently; but (3) it cannot solve other types of problems—i.e. problems which are not diagnostic in nature—that the deep knowledge structure potentially could handle.

For the past several years, we have been developing a medical diagnosis system called MDX (Mittal, Chandrasekaran & Smith, 1979; Gomez & Chandrasekaran, 1981). This system embodies the above thesis. An attempt has been made to compile in the knowledge structure of MDX all the relevant aspects of deep knowledge available to our human experts. Before we argue for the adequacy of such a compiled structure, it will be useful to give a brief characterization of the type of diagnostic problems that MDX deals with, and how it solves them.

2. MDX and the diagnostic task

The diagnostic process is, in general, complex, but a core task in this process—a task we will call "diagnostic" in this paper with some loss of generality—is the *classification* of a case description as a node in a pre-determined diagnostic hierarchy.† For the purpose of current discussion let us assume that all the data that can be obtained are already there, i.e. the additional problem of launching exploratory procedures such as ordering new tests, etc., does not exist. The following brief account is a summary of the more detailed account given in Gomez & Chandrasekaran (1981) of diagnostic problem-solving.

Let us imagine that corresponding to each node of the classification hierarchy alluded to earlier we identify a "concept". More general classificatory concepts are higher in the structure, while more particular ones are lower in the hierarchy. The total diagnostic knowledge is then distributed through the conceptual nodes of the hierarchy in a specific manner to be discussed shortly. In the medical domain, a fragment of such a hierarchy might be:

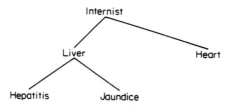

FIG. 1. Example diagnostic structure.

The problem-solving for this task will be performed top-down, i.e. the topmost concept will first get control of the case, then control will pass to an appropriate successor concept, and so on. In this case, INTERNIST first establishes that there is in fact a disease, then LIVER establishes that the case at hand involves some liver disease, while HEART will reject the case as not being in its domain. After this, CHOLESTASIS may establish itself and so on.

Each of the concepts in the classification hierarchy contains "how-to" knowledge represented as a collection of *diagnostic rules*. These rules are of the form: ⟨symptoms⟩ → evidence about ⟨concept in hierarchy⟩, e.g. "If high SGOT, add n units of evidence in favor of cholestasis". Because of the fact that when a concept rules itself out from relevance to a case, all its successors also get ruled out, large portions of the diagnostic knowledge structure never get exercised. On the other hand, when a concept is properly invoked, a small, highly relevant set of rules comes into play.

The problem-solving regime that is implicit in the structure can be characterized as an *establish–refine* type. That is, each concept first tries to establish or reject itself. If it succeeds in establishing itself, then the refinement process consists of checking which of *its* successors can establish themselves. Each concept has several clusters of rules, primarily consisting of confirmatory rules, and exclusionary rules. The evidence for confirmation and exclusion can be suitably weighted and combined to arrive at a conclusion to establish, reject, or suspend it. The last-mentioned situation may arise if there is insufficient data to make a decision.

† The issue of whether there is always a hierarchical classification structure is often a source of debate. We refer the reader to Chandrasekaran (1983a) for a brief discussion of this issue. For the sort of plausibility arguments that we advance in this paper, whether the classification structure is precisely a hierarchy, etc., does not matter.

The concepts in the hierarchy are clearly not static bodies of knowledge. They are active in problem-solving. But they only have knowledge about establishing and rejecting the relevance of that conceptual entity. Thus, they may be termed "specialists", in particular, "diagnostic specialists".

The above account of diagnostic problem-solving is quite incomplete. For example, we have not indicated how multiple diseases can be handled within the framework above, in particular when a patient has a disease secondary to another disease. There are important issues regarding the conditions under which the top-down control may need to be locally modified. A more powerful model for problem solving is outlined in Gomez & Chandrasekaran (1981). For the purposes of this paper, however, this simpler account is sufficient.

3. Example analysis

3.1. REASONING WITH DEEP KNOWLEDGE

Before we get into the example proper, the following background will be useful. (In this entire discussion, the interests of clarity have overridden the interests of completeness and medical accuracy. Much of the discussion is simplified to make the essential technical points.) The MDX system diagnoses in a syndrome called Cholestasis, which is a condition caused when the secretion of bile from the liver or its flow to the duodenum is blocked. Such a blockage may have a number of causes. MDX attempts to pinpoint the cause in a given case. A subset of causes can be grouped under the category "Extrahepatic Obstruction (EHO)", i.e. blockage due to an obstruction of bile flow outside the liver. In this example, we will assume that the physician has established that a cholestatic condition is present, and he is examining whether the cause is extra-hepatic, in particular which of a number of possible extra-hepatic causes may be the underlying reason. Bile flows from the liver into the duodenum via the bile duct. The following is a sequence of explicit reasoning steps by a (hypothetical) physician who does not yet have a compiled structure for diagnostic reasoning in this subdomain.

1. The bile duct is a flexible and somewhat elastic tube. If such a tube has a blockage at some point and if there is fluid pressure building up, then it will be dilated on the "upstream" side of the blockage. Thus, if there is an extra-hepatic obstruction, the biliary tree inside the liver and a portion outside should be dilated. This should be visible in various imaging procedures as a specific visual pattern, call it ⟨pattern 1⟩. Thus, EHO causes ⟨pattern 1⟩ in X-ray of the region.

2. Given a flexible duct, obstruction can be caused because there is a physical object in the duct, a contraction such as a stricture is present, some object outside the duct is pressing on the duct, or the internal diameter of the duct is reduced for some reason. Physical objects that I can think of in this context are biliary stone and a tumor in the duct. Looking at the anatomy of the region, nearby organs are gall bladder and pancreas. Cancers in these organs can press on the ducts and cause obstruction.

3. Biliary stones show up as a characteristic pattern (call it ⟨pattern 2⟩) in a cholangiogram. Since the stones cause obstruction and an increase in the peristaltic action of the duct, they can cause acute colicky abdominal pain.

4. Cholangitis (inflammation of the bile duct) can cause swelling, and reduce the duct diameter and bile flow. A stricture can be caused by the surgical wound during a prior bile duct surgery not healing properly. . . .

3.2. THE USE OF DIFFERENT LEVELS OF KNOWLEDGE

In step 1 of the sequence in section 3.1, very general knowledge about flexible pipes as well as knowledge about anatomy of the region and about imaging procedures is accessed, and from all these, a highly specific rule is compiled relating a diagnostic state to certain X-ray patterns. In step 2, again very general knowledge about flexible ducts is used in conjunction with an anatomical model of the region and other causal knowledge to generate a list of possible causes. In step 3, knowledge about imaging procedures, and about the physical properties of stones and ducts, is ·used to *infer certain consequences of the diagnostic state being present.* Clearly, a variety of knowledge structures are used in the above reasoning fragment, many of them surely deserving the name "deep" or "underlying" knowledge. (After all, what is deeper during medical diagnosis than very general knowledge about flexible ducts?) Note also that not all these pieces of knowledge are at the same level: some had to do with flexible ducts, while others were highly domain-dependent pieces such as knowledge about cholangiograms. In this particular instance the physician reasoned about the relationship between colicky abdominal pain and stones from more basic pieces of knowledge about peristaltic ducts. A more experienced physician may simply use a piece of knowledge, "biliary stones→colicky abdominal pain". Thus any such reasoning will be a mixture of such compiled pieces and more general items of knowledge.

3.3. DIAGNOSTIC STRUCTURE

Now we shall attempt to show that we can create a diagnostic structure as a result of the above reasoning. Once this structure is available, most of the steps in the above reasoning can be skipped.

Consider the hierarchy of specialists in Fig. 2. Let us refer to this structure as the D structure for convenience.

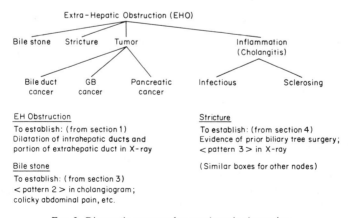

FIG. 2. Diagnostic structure for extrahepatic obstruction.

When a case is established to be cholestatic, control will pass to EHO. If EHO is able to establish itself (i.e. if intrahepatic and a portion of the extrahepatic ducts are dilated—see establish rule in Fig. 2 for EHO), it will call its successors (in parallel, for purposes of current discussion). Each successor will similarly establish or reject itself by using appropriate rules. Typically only one of them will be established at each level. The one(s) that are established refine the hypothesis by calling their successors. The tip nodes that are established provide the most detailed diagnostic classification of the case. (Again we caution that this portion of D is for illustration only; the reality is much more complicated.)

If one were to denote by U (standing for underlying knowledge) all the knowledge structures that were accessed during the physician's reasoning described above, and by D the knowledge in the above diagnostic structure (Fig. 2), then the claim is that *all the knowledge in U that plays a role in diagnosis is in D in a form that is directly usable. If D fails to solve a problem, a resort to U will not improve the situation.* [An exception, which does not counter the basic argument, but may have some implementation consequences is discussed in point (5) of section 4]. Section 4 of the paper is essentially devoted to arguing for this claim in some detail.

3.4. RELATION BETWEEN D AND U STRUCTURES

In structure D of Fig. 2, several diagnostic states are identified and organized hierarchically. (The justification for this particular hierarchy of diagnostic states as opposed to other alternatives cannot be provided by reference only to knowledge in U. For example, one might have chosen to place biliary duct tumor and biliary stone as children of a node called physical obstruction, rather than group bile duct tumor with gall bladder cancer as we have done. Meta-criteria such as therapeutic implications or commonality of knowledge needed for establishing the states come into play. For example, many of the manifestations of cancer are the same whether it is bile duct cancer or GB cancer; hence the choice presented. While further discussion of this issue is beyond the scope of this paper, the point to be noted is that D is organized for efficiency in the diagnostic process, a consideration which may not be directly available from U.) The complex reasoning in step 1 of section 3.1, based on anatomy and knowledge about ducts, is simply compiled into the establishment rule in the EHO specialist. Step 2 resulted in candidates for further specializations of the hypothesis about the cause of cholestasis. Step 3 resulted in the procedure for establishing biliary stone as cause of cholestasis. The causal reasoning in step 4 resulted in a similar procedure for stricture.

4. The adequacy of D structures for diagnosis

Let us briefly look at some of the ways in which D may fail to solve a case.

(1) A concept is missing in D because a needed chunk of knowledge is missing in U. For example, suppose in a particular case EHO is established, and evidence of inflammation is present. Suppose further that cholangitis is ruled out. D will be able to provide only an incomplete solution to the problem. It turns out that there is another inflammation possibility in that region that can also cause obstruction: pancreatitis. This piece of knowledge was missing in U. This resulted in D

missing a concept, a sibling of Cholangitis. Thus the problem can be traced to an inadequacy of U. An automated diagnostic system endowed with the same U will gain no further ability to solve this case by referring to U, deeper though it is.

(2) Some establish/reject knowledge is missing in a specialist. Again such a deficiency can be traced to missing chunk of knowledge in U.

(3) D's problem-solving (PS) strategy is not powerful enough. [As indicated in section 2 the problem-solving strategy outlined there is inadequate, and a more powerful strategy is given in Gomez & Chandrasekaran (1981).] In this case a reference to U still will not be able to do the trick. The weakness of the PS strategy of D means that the system designer has not fully comprehended the *use* of knowledge for diagnostic problem-solving. Thus access to U will not help, since the issue in question is precisely how to properly use the knowledge. A system which knows how to use, the knowledge in U can equally be rewritten to embed that improved use of knowledge in the problem-solving strategy of D. For this reason the arguments for compilation are not strongly dependent on the correctness of the MDX approach in detail.

(4) The knowledge is there in U, D's PS strategy is powerful enough, but the knowledge was improperly compiled in D. This in fact often happens. But again access to U during the problem-solving by D will be useless, since the ability to use the knowledge in U effectively for diagnosis implies the ability to identify the proper use of that knowledge for compilation into D. In fact we feel the problem of automatic generation of D from U is a much harder research problem. In our research group we have initiated work on this problem. The organization of knowledge in U should support generation of D by domain-independent diagnostic structure building procedures.

(5) U has the relevant knowledge, but compiling all the relevant ways in which that knowledge can come into play in diagnosis will result in a combinatorial problem in representing them in D. A specific example will motivate this situation. In a Clinico-Pathological Conference (CPC) case involving EHO that MDX was presented with, the system established "inflammation" (see Fig. 2), but could not establish any of its successors. It turned out that in an unusual sequence of events a piece of fecal matter from the intestine had entered the bile duct through the Sphincter of Oddi, and was physically obstructing bile flow. (This possibility escaped not only MDX, but all of the physicians who analyzed the case in the CPC. It was discovered during surgery.) One could suppose that in theory all such possibilities could be compiled but, even if they could be, the number of possibilities in which some piece of matter from one part of the body could end up through some fistula (say) in the bile duct is quite large. One way out in this kind of situation is for D to simply have a procedure which calls the underlying knowledge structure U at this juncture in its reasoning. But notice that all the possibilities that were not compiled in D because they would be too numerous will nevertheless all have to be generated at this point. Thus in principle it is simply a run-time compilation of the needed portion of D.† In fact precisely because of the combinatorial nature of this situation, all the physicians, endowed with quite a powerful U-structure, failed to think of the real cause.

---

† Strictly speaking, it is possible that the deep structure may have knowledge that might enable this generation process to be efficiently pruned; this knowledge, not being diagnostic in our sense, may not be appropriate for compilation into D.

The situation corresponding to the foregoing discussion in (5) can be quite important in the implementation of expert systems. One might choose to access carefully circumscribed portions of U at certain points in diagnostic reasoning, along with the compilation procedure, rather than store the results of compilation of that fragment in D, in order to save storage space. But this doesn't change the essence of the argument.

D may include a data base of patterns for further efficiency. Some of the major specialists such as EHO may contain a small set of patterns which can be used to dispose of some common possibilities without using the full problem-solving capabilities of D. For example, assume for the sake of the argument that EHO due to bile stone accounts for 90% of all EHO cases, and that due to stricture for a further 5%. (This happens not to be true, thus we re-emphasize that this is purely for making a point.) Now, the EHO specialist may, after establishing itself, access a small data base of patterns that may help to establish or dispose of these possibilities with a small effort. Thus the system can solve 95% of the cases with extreme efficiency. But the important point is that if this data base of patterns is exhausted before problem solution, the rest of D is still available for more thorough problem-solving.

5. Interaction between different compiled structures

Almost all of the discussion in the paper so far has dealt with the generic task of diagnosis. We have argued elsewhere (Chandrasekaran, 1983a) that there exist other generic tasks, each with a particular way of using knowledge for problem-solving. Corresponding to each such task in a domain, one may construct a compiled problem-solving structure. In the paper mentioned above, we have identified a few other generic tasks that we have encountered in the analysis of medical reasoning. While most of the arguments in this paper do not depend upon the correctness of the details of the MDX approach to diagnostic problem-solving, they do implicitly depend upon the existence of diagnosis as a generic task. This is what makes it possible to compile the underlying knowledge in a structure such as D. In our view, identification of further such generic tasks and how these structures interact in complex probem-solving situations is an important challenge for AI research.

Interaction between compiled structures is important to the current discussion for the following reason. While we have argued in section 4 for the adequacy of the D structure for diagnosis we have not touched on a potential source of combinatorial growth in the knowledge needed in D. Example: Consider a situation where D has the following rule as part of a specialist: "If exposure to anesthetics, consider hepatitis" (we assume that this specialist is a subspecialist of Liver, which is Established). Suppose the physician notices no entry in the data base regarding anesthetics exposure, but sees evidence of recent major surgery, he should infer anesthetic exposure, and thus consider hepatitis. But this reasoning, i.e. deduction from surgery to anesthetic exposure, is not diagnostic reasoning, not being a ⟨finding⟩ → ⟨diagnostic hypothesis⟩ type. One way for the D structure to handle this correctly is to have another compiled rule: "If evidence of major surgery, consider hepatitis"; i.e. the deduction "surgery → anesthetic exposure" is not made during problem-solving time, but implicit in the collection of rules. This approach, however, has the potential for combinatorial explosion in the number of rules in the diagnostic specialists due to its need for explicitly encoding all such finding-to-finding deductions.

There is an alternative. Instead of viewing diagnostic problem solving as being performed either with knowledge structure U or with structure D, we identify a knowledge and problem-solving structure devoted exclusively to the finding-to-finding reasoning discussed above, where the knowledge that is relevant to this type of reasoning is compiled. The type of reasoning that this structure performs can be characterized variously as a form of knowledge-based information retrieval or associative memory. In this broader framework, D has compiled in it only direct diagnostic knowledge. In the above example involving anesthetics, e.g. it will only have the basic compiled diagnostic rule "anesthetics → hepatitis", and will turn to the information retrieval structure for confirming or denying the exposure to anesthetics. We have provided in Mittal & Chandrasekaran (1981) an account of the nature of knowledge organization and problem-solving that underlies this kind of information retrieval activity. Expert reasoning is viewed in this framework at least partly as a cooperative effort between different compiled structures. In Chandrasekaran & Mittal (1983) we have given a detailed discussion of how medical diagnosis is done in the MDX system by interaction with other specialist structures.

6. On causal knowledge as a deep model

While it is not our intention in this paper to provide a theory of what sorts of representations qualify as appropriate deep models for expert systems, it seems useful to examine a proposal that is often made that such models should represent "causal knowledge". Systems such as CASNET (Weiss, Kulikowski, Safir & Amarel, 1978), ABEL (Patil, 1981) and CADUCEUS (Pople, 1982) have knowledge which is explicitly a set of cause–effect relations. But whether using a collection of cause–effect relations as its knowledge is sufficient for a system to qualify as a deep model is debatable. Any diagnostic system, be it a surface system or a deep one, is able to relate findings to diagnostic hypotheses only because of the causal relationship that exists between the disease and the finding. Whether the documentation of the system refers to its knowledge as having the form "⟨disease⟩ causes ⟨finding⟩", or the form "⟨finding⟩ gives evidence of ⟨disease⟩" is not the important issue. Any potential for deep reasoning must come from some other property of a system than that it uses causal knowledge.

Perhaps what is at the back of the intuition that causal reasoning is important in a deep model is the frequent need to establish causal connections at different levels of detail. That is, in addition to (or instead of) causal knowledge that relates diagnostic hypotheses (at the level of decision-making interest) to observable findings, a system may have knowledge that relates diagnostic hypotheses to intermediate pathophysiological (pp) states, pp states to other pp states, and pp states to findings. Such a system may be able, using this knowledge, to ensure that a conclusion from findings to diagnostic decisions can be justified by generating a plausible sequence of intermediate pp states. Knowledge at the increased level of detail also appears to give an additional "explanation" capability.

Two related points seem worth making here. First, the semantics of "cause" implicit in the use of causal knowledge by systems using such cause–effect relations at different levels of detail are no different from that used by systems such as MYCIN or MDX. As far as the use of knowledge is concerned, nothing would change in the so-called causal model systems if the knowledge "A causes B" were restated as "If B, then infer A". These systems have no previleged understanding of causation. The second point is that the cause–effect relations in these systems are *compiled* ones, i.e. the reasoning mechanisms in these systems do not *generate* the cause–effect information from some model of the underlying reality. There is an intuitive expectation that understanding a domain gives one an ability to generate such relationships.

Some recent work is oriented towards giving expert systems such a capability. deKleer & Brown (1981) on functional models of devices and Kuipers (1982) on deriving behavior from structure are examples of such approaches.

7. Providing explanations

The idea that deep knowledge structures are needed for providing explanations has often been proposed. It is perhaps worth examining the concept of explanation itself. Referring to the structure D in Fig. 2, suppose the diagnostic conclusion in a particular case was "EHO due to bile stone". Suppose we ask

Q1. Why do you conclude that this patient has EHO due to bile stone?

The system at that point may refer to its procedure for establishing that hypothesis and reply "Dilatation of intrahepatic and a portion of extrahepatic bile duct was observed in X-ray, thus I concluded EHO. Because further a characteristic pattern, ⟨pattern 2⟩ was observed in Cholangiogram, and because colicky abdominal pain was observed, I concluded bile stones". Note that the system is simply telling which of the rules was found to match for each specialist that was established. Some people may accept this explanation. At this level D can be quite satisfactory. Suppose a further question was asked:

Q2. "Why does a dilated IH bile duct indicate EHO?"

D, as it stands, cannot answer this question, since the rationale for the rule was left behind in U, at the time of compilation. But note that giving an explanation for this question does not need any additional problem-solving. We can simply associate with each piece of compiled knowledge in D a text string that explains its rationale. This is possible because the answer to Q1 is patient-specific and thus requires problem-solving with D, while the answer to Q2 is not patient-specific, and has already been generated in designing D.

Since in a human expert D is built by learning and reasoning processes from U, inadequacies in D are often correlated with inadequacies in U. That is why we are suspicious of physicians who we feel cannot explain the basis of their conclusions by reference to deeper knowledge structures. Since U changes with time given new discoveries in medicine, most of us as patients would like to assume that our physicians are capable of translating the relevant changes in U to appropriate changes in D. An ability to explain by calling forth appropriate fragments of U is our assurance that the expertise that the physician is using to diagnose our illness is built on the basis of appropriate learning processes. Thus a system such as D, when it answers Q1, is giving evidence of the correctness of its own problem-solving, while its answer to Q2 is evidence of the correctness of those human experts who provided the underlying knowledge structures and the compilation process. The fact that the compiled structures of human experts are often incomplete adds to the need for assurances that reasonable deep models are present, since the latter would be needed for those cases which require the missing knowledge in D.

8. Concluding remarks

One aspect of our analysis has been to examine more carefully what is generally viewed as a simple "deep" vs "surface" distinction in the level of reasoning in expert systems. Our argument in this paper, and our work in general, is oriented towards a better understanding of the territory that lies in between. We can map this territory by listing several knowledge structures at differing levels of compilation as follows.

Table look-up. A situation is directly recognized as one for which the problem-solver has a solution. In most fields of expertise, experts have a collection of such ready-to-use pieces of knowledge. In general this approach is combinatorially too large for all of the domain. Typically, experts have such stored patterns to account for the most frequent or most important situations.

Partial pattern-matching. This is the approach used in several first-generation expert systems such as MYCIN. Instead of the totality of the problem-solving situation being recognized as one for which the problem-solver has a solution, portions of the situation are recognized and partial solutions are collected by using simple rules. The formulation of partial solutions may enable further portions of the situation to be recognized, and often the total solution can be synthesized after several such cycles. Depending on the origin of the "pattern → action" knowledge such a system may be more-or-less *ad hoc*. [In Chandrasekaran (1983*b*) we discuss the different types of rules in rule-based systems.] However, there are several problems with this partial pattern-matching approach. When a number of patterns match, there may arise problems of focus in reasoning, viz. which line of thought to pursue.

Compiled structures for generic problem-solving types. This is the approach that we have been elaborating in our work on MDX, and one we have argued for in this paper. Here, basically deeper knowledge of the domain is analyzed from the view point of the role that knowledge plays in problem-solving of those generic types, and this knowledge is compiled in structures specializing and tuned for that type of problem-solving. We have provided in this paper a set of plausibility arguments for how such a compiled diagnostic structure may be able to solve a variety of diagnostic problems as well as any deep model of the domain can. We have also suggested that such compiled structures may not be subject to combinatorial explosion in general, since only those aspects of the deep knowledge that are appropriate to problems of that type are compiled, and there are strong organizational principles that govern the formation of these structures. Further, by letting structures specializing in different types of problem-solving interact, the potential for combinatorial growth in each of the structures is minimized.

The term "compiled" in describing these structures needs to be properly understood not as compiling solutions to problems as in the table look-up method above, but as compiling knowledge in a form ready to be used for a class of problems of a given type.

"Deep" structures. We have specifically avoided in this paper any speculation so far on what this form of knowledge is, since that is not relevant to our main purpose. We argue in section 6 that the intuition that it has to be "causal" is not sufficient.

Whatever the structure is, it ought to be able to support compilation processes for different types of generic problem-solving types.

The approach we are arguing for sharpens some of the questions surrounding how a novice learns to become an expert. We suggest that this issue is facilitated by having some idea of what are the target structures that are being learned. In the case of diagnosis, we suggest that learning to become an expert diagnostician is the process of compiling a structure such as D. Thus research into what sorts of knowledge structures U consists of and what sorts of processes are powerful enough to produce problem-solving structures such as D will significantly advance our ability to produce learning systems of considerable importance to expert system design.

We thank Jack Smith, M.D., Michael Coombs and John Josephson for helpful comments on an earlier version of the paper. We acknowledge the contributions of Fernando Gomez to the approach outlined here.

The preparation of this paper was supported by NSF grant MCS-8103480. The computing resources of Rutgers University Laboratory for Computer Science Research were used in system development, and this was made possible by the Biotechnology Resources Division of NIH, under grant RR-00643.

References

CHANDRASEKARAN, B. (1983a). Towards a taxonomy of problem solving types. *AI Magazine*, **4** (1), 9–17.

CHANDRASEKARAN, B. (1983b). Expert systems: matching techniques to tasks. In REITMAN, W., Ed., *Artificial Intelligence Applications for Business*, Norwood, New Jersey: Ablex Publishing (forthcoming).

CHANDRASEKARAN, B. & MITTAL, S. (1983). Conceptual representation of medical knowledge for diagnosis by computer: MDX and related systems. In YOVITS, M. C., Ed., *Advances in Computers*, vol. 22, pp. 217–293. New York: Academic Press.

DeKLEER, J. & BROWN, J. S. (1981). Mental models of physical mechanisms and their acquisition. In ANDERSON, J. R., Ed., *Cognitive Skills and Their Acquisition*, pp. 285–309. Hillsdale, New Jersey: Erlbaum Associates.

GOMEZ, F. & CHANDRASEKARAN, B. (1981). Knowledge organization and distribution for medical diagnosis. *IEEE Transactions on Systems, Man, and Cybernetics*, **SMC-11** (1), 34–42.

HART, P. E. (1982). Direction for AI in the eighties. *SIGART Newsletter*, **79**, 79 (January).

KUIPERS, B. (1982). Commonsense reasoning about causality: deriving behavior from structure. *Working Papers in Cognitive Science*, # 18. Tufts University, Medford, Massachusetts.

MICHIE, D. (1982). High-road and low-road programs. *AI Magazine*, **3** (1), 21–22.

MITTAL, S. & CHANDRASEKARAN, B. (1981). Software design of knowledge-directed database systems. In *Proceedings First Conference on Foundations of Software Technology and Theoretical Computer Science*, Bombay, India, December 1981.

MITTAL, S., CHANDRASEKARAN, B. & SMITH, J. (1979). Overview of MDX—a system for medical diagnosis. In *Proceedings Third Symposium Computer Applications in Medical Care*, Washington, D.C., October 1979, pp. 34–46.

PATIL, R. S. (1981). Causal representation of patient illness for electrolyte and acid–base diagnosis. *Ph.D. Dissertation*, *TR-267*. M.I.T. Laboratory for Computer Science, Cambridge, Massachusetts.

POPLE, H. E. (1982). Heuristic methods for imposing structure on ill-structured problems. In SZOLOVITS, P., Ed., *Artificial Intelligence in Medicine*, pp. 119–190. Boulder, Colorado: Westview Press.

SHORTLIFFE, E. H. (1976). *Computer-based Medical Consultations: MYCIN*. New York: Elsevier/North-Holland Inc.

WEISS, S. M., KULIKOWSKI, C. A., AMAREL, S. & SAFIR, A. (1978). A model-based method for computer-aided medical decision-making. *Artificial Intelligence*, **11**, 145–172.

Diagnostic expert systems based on a set covering model

JAMES A. REGGIA,*† DANA S. NAU† AND PEARL Y. WANG§

*Departments of *Neurology, †Computer Science and §Mathematics, University of Maryland*

This paper proposes that a generalization of the set covering problem can be used as an intuitively plausible model for diagnostic problem solving. Such a model is potentially useful as a basis for expert systems in that it provides a solution to the difficult problem of multiple simultaneous disorders. We briefly introduce the theoretical model and then illustrate its application in diagnostic expert systems. Several challenging issues arise in adopting the set covering model to real-world problems, and these are also discussed along with the solutions we have adopted.

1. Introduction

A diagnostic problem can be defined to be a problem in which one is given a set of abnormal findings (manifestations) for some system, and must explain why those findings are present. Problems of this kind are very common: they include diagnosing a patient's signs and symptoms, determining why a computer program failed, deciding why an automobile will not start, finding the cause of noise in a plumbing system, localizing a fault in an electronic circuit, etc. Because of this ubiquity, developing general methods for expert systems which support the decision making of human diagnosticians is an important issue at present.

This paper introduces a new model for diagnostic expert systems based on the concept of minimal set covers. This model is of interest because it captures several intuitively plausible features of human diagnostic inference, it directly addresses the problem of multiple simultaneous causative disorders, and it provides a basis for a theory of diagnostic inference.

In the following, section 2 discusses the set covering model, and section 3 explains how the model can be adopted for use in expert diagnostic systems. Section 4 and Appendix B give examples of operational expert systems based on set covering, and section 5 presents in a more detailed fashion some of the issues involved in implementing these systems. Section 6 contains some concluding remarks.

2. The set covering model

In the set covering model the underlying knowledge for a diagnostic problem is organized as pictured in Fig. 1(a) (a table of symbols is given in Appendix A). There are two discrete finite sets which define the scope of diagnostic problems: **D**, representing all possible *disorders* d_i that can occur, and **M**, representing all possible *manifestations* m_i that may occur when one or more disorders are present. For example, in medicine, **D** might represent all known diseases (or some relevant subset of all diseases,

All correspondence to: Dr James Reggia, Department of Neurology, University of Maryland Hospital, Baltimore, Maryland 21201, U.S.A.

35

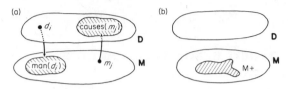

FIG. 1. Organization of diagnostic knowledge (a) and problems (b).

see below), and **M** would then represent all possible symptoms, examination findings, and abnormal laboratory results that can be caused by diseases in **D**. We will assume that $\mathbf{D} \cap \mathbf{M} = \varnothing$.

To capture the intuitive notion of causation, we assume knowledge of a relation $\mathbf{C} \subseteq \mathbf{D} \times \mathbf{M}$, where $\langle d_i, m_j \rangle \in \mathbf{C}$ represents "d_i can cause m_j". Note that $\langle d_i, m_j \rangle \in \mathbf{C}$ does not imply that m_j always occurs when d_i is present, but only that m_j may occur. For example, a patient with a heart attack may have chest pain, numbness in the left arm, loss of consciousness, or any of several other symptoms, but none of these symptoms are necessarily present.

Given **D**, **M**, and **C**, the following sets can be defined:

$$\mathrm{man}(d_i) = \{m_j | \langle d_i, m_j \rangle \in \mathbf{C}\} \qquad \forall d_i \in \mathbf{D}, \text{ and}$$

$$\mathrm{causes}(m_j) = \{d_i | \langle d_i, m_j \rangle \in \mathbf{C}\} \qquad \forall m_j \in \mathbf{M}.$$

These sets are depicted in Fig. 1(a), and represent all possible manifestations caused by d_i, and all possible disorders that cause m_j, respectively. These concepts are intuitively familiar to the human diagnostician. For example, medical textbooks frequently have descriptions of diseases which include, among other facts, the set $\mathrm{man}(d_i)$ for each disease d_i. Physicians often refer to the "differential diagnosis" of a symptom, which corresponds to the set $\mathrm{causes}(m_j)$. Clearly, if $\mathrm{man}(d_i)$ is known for every disorder d_i, or if $\mathrm{causes}(m_j)$ is known for every manifestation m_j, then the causal relation **C** is completely determined. We will use $\mathrm{man}(\mathrm{D}) = \bigcup_{d_i \in \mathrm{D}} \mathrm{man}(d_i)$ and $\mathrm{causes}(\mathrm{M}) = \bigcup_{m_j \in \mathrm{M}} \mathrm{causes}(m_j)$ to indicate all possible manifestations of a set of disorders D and all possible causes of any manifestation in M, respectively.

Finally, there is a distinguished set $\mathbf{M}^+ \subseteq \mathbf{M}$ which represents those manifestations which are known to be present (see Fig. 1(b)). Whereas **D**, **M**, and **C** are general knowledge about a class of diagnostic problems, $\mathbf{M}^+$ represents the manifestations occurring in a specific case.

Using this terminology, we can now make the following definition.

Definition. A *diagnostic problem P* is a 4-tuple $\langle \mathbf{D}, \mathbf{M}, \mathbf{C}, \mathbf{M}^+ \rangle$ where these components are as described above.

We will assume in what follows that diagnostic problems are well-formed in the sense that $\mathrm{man}(d_i)$ and $\mathrm{causes}(m_j)$ are always non-empty sets.

Having characterized a diagnostic problem in these terms, we now turn to defining the solution to a diagnostic problem by first introducing the concept of explanation.

Definition. For any diagnostic problem P, $\mathrm{E} \subseteq \mathbf{D}$ is an *explanation* for $\mathbf{M}^+$ if; (i) $\mathbf{M}^+ \subseteq \mathrm{man}(\mathrm{E})$, or in words: E *covers* $\mathbf{M}^+$; and (ii) $|\mathrm{E}| \leq |\mathrm{D}|$ for any other cover D of $\mathbf{M}^+$, i.e. E is *minimal*.

This definition captures what one intuitively means by "explaining" the presence of a set of manifestations. Part (i) specifies the reasonable constraint that a set of disorders E must be able to cause all known manifestations M^+ in order to be considered an explanation for those manifestations. However, that is not enough: part (ii) specifies that E must also be one of the smallest sets to do so. Part (ii) reflects the Principle of Parsimony or Ockham's Razor: the simplest explanation is the preferable one. This principle is generally accepted as valid by human diagnosticians. Here, we have equated "simplicity" with minimal cardinality, reflecting an underlying assumption that the occurrence of one disorder d_i is independent of the occurrence of another.

With these concepts in mind, we can now define the solution to a diagnostic problem.

Definition. The *solution* to a diagnostic problem P, designated Sol(P), is the set of all explanations for M^+.

The concepts defined above are illustrated in the following example.

Example. Let $P = \langle \mathbf{D}, \mathbf{M}, \mathbf{C}, M^+ \rangle$ where $\mathbf{D} = \{d_1, d_2, \ldots, d_9\}$, $\mathbf{M} = \{m_1, \ldots, m_6\}$, and man($d_i$) and causes($m_j$) are as specified in Table 1. Note that the top (or bottom) half of Table 1 implicitly defines the relation $\mathbf{C}$, because $\mathbf{C} = \{\langle d_i, m_j \rangle | m_j \in \text{man}(d_i) \text{ for some } d_i\}$. Let $M^+ = \{m_1, m_4, m_5\}$. Note that no single disorder can cover (account for) all of M^+, but that some pairs of disorders do cover M'. For instance, if $D = \{d_1, d_7\}$ then $M^+ \subseteq \text{man}(D)$. Since there are no covers for M^+ of smaller cardinality than D, it follows that D is an explanation for M^+. Careful examination of Table 1 should convince the

TABLE 1

Knowledge about a class of diagnostic problems. The relation C is implicitly defined by either the top or bottom half of this table

d_i	man(d_i)
d_1	$m_1\, m_4$
d_2	$m_1\, m_3\, m_4$
d_3	$m_1\, m_3$
d_4	$m_1\, m_6$
d_5	$m_2\, m_3\, m_4$
d_6	$m_2\, m_3$
d_7	$m_2\, m_5$
d_8	$m_4\, m_5\, m_6$
d_9	$m_2\, m_5$

m_j	causes(m_j)
m_1	$d_1\, d_2\, d_3\, d_4$
m_2	$d_5\, d_6\, d_7\, d_9$
m_3	$d_2\, d_3\, d_5\, d_6$
m_4	$d_1\, d_2\, d_5\, d_8$
m_5	$d_7\, d_8\, d_9$
m_6	$d_4\, d_8$

reader that

$$\text{Sol(P)} = \{\{d_1\, d_7\}\, \{d_1\, d_8\}\, \{d_1\, d_9\}\, \{d_2\, d_7\}\, \{d_2\, d_8\}\, \{d_2\, d_9\}\, \{d_3\, d_8\}\, \{d_4\, d_8\}\}$$

is the set of all explanations for M^+.

It is of interest to compare the model of diagnostic problems presented here with the classic set covering problem. The set covering problem is typically stated along the following lines (Edwards, 1962):

> For a finite set S of elements and a family F of subsets of S, a cover K of S from F is a sub-family $K \subseteq F$ such that $\bigcup(K) = S$. A cover K is called minimum if its cardinality is as small as possible.

In this definition, S corresponds to M^+ and F corresponds to **D** in the sense that each $d_i \in \mathbf{D}$ labels a subset of M^+ (the intersection of man(d_i) with M^+). A minimum cover K corresponds roughly to the idea of an explanation E except man(E) is required only to contain M^+ rather than be equal to M^+.

3. Expert systems using the set covering model

We now turn to the description of expert systems for diagnostic problem solving based on the set covering model presented above. Such systems are organized as shown in Fig. 2 and consist of three parts.

1. A *database*, which is divided into case-specific information and general knowledge about some domain of diagnostic problems. We will use the term *knowledge base* for the latter.

2. An *inference mechanism* which is a hypothesize-and-test process that mimics diagnostic reasoning by using the set covering model.

3. A *user interface* which accepts assertions and queries from the user and translates them into internal data structures.

We now present a specific example of an expert system called System D for diagnostic problem solving. This implemented system illustrates how the set cover model can be adopted to the demands of real world problems. While System D is medically oriented, it should be remembered that the set cover model is domain-independent and not restricted to problems of medical diagnosis.

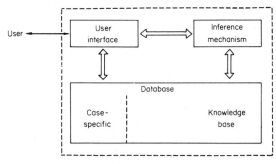

FIG. 2. Architecture of an expert system based on the set covering model.

System D is a relatively large expert system for diagnosing patients with dizziness. Dizziness is in general a very difficult diagnostic problem for the physician because there are numerous potential causes that are distributed across multiple medical specialties. Examples of possible diagnoses include:

orthostatic hypotension secondary to drugs (orthostatic hypotension is a fall in blood pressure upon standing up, and can be a side effect of certain medications);
heart disease, such as an irregular heart beat or an abnormal heart valve;
basilar migraine: headache due to painfully dilated blood vessels which supply blood to the balance centers of the brain;
inner ear diseases: these interfere with the balance mechanisms of the inner ear, and include viral labyrinthitis, Meniere's disease, and otosclerosis; and
hyperventilation: overbreathing, typically secondary to anxiety.

It is entirely possible that more than one cause of dizziness could be present simultaneously.

The knowledge base for System D is derived from numerous references and currently contains information about 50 causes of dizziness. It was built using KMS, a domain-independent software facility for constructing expert systems (Reggia, 1981). We now describe its components in detail.

THE CASE-SPECIFIC DATABASE

The case-specific database for System D contains a collection of assertions that describe a specific diagnostic problem. For example,

        AGE=50;

        DIZZINESS=PRESENT
           [TYPE=VERTIGO; COURSE=EPISODIC]; and

        NEUROLOGICAL SYMPTOMS=DIPLOPIA

represent three assertions that might appear in the database. Each assertion is of the form

        attribute relation value [elaboration],

so the three statements here mean: "A 50-year-old individual with episodic vertigo (a type of dizziness where one feels a sensation of motion) and double vision (diplopia)". During a problem solving session this case-specific information is acquired in a sequential fashion, generally in response to questions generated by the expert system. The legal attributes and their possible values are predefined in a database schema by the creator of the expert system (the "knowledge base author").

THE KNOWLEDGE BASE: REPRESENTING DIAGNOSTIC KNOWLEDGE

One of the attractive features of the set cover model is that it permits the organization of diagnostic knowledge in a form familiar to the human diagnostician. Information in the knowledge base is organized into frame-like entities called DESCRIPTIONs. Each DESCRIPTION provides a textbook-like summary of the disorder with which it is associated. An example of a DESCRIPTION from System D is illustrated in Fig. 3. To understand this descriptive knowledge representation more fully, it is necessary

```
MENIERE'S DISEASE  <L>
  [DESCRIPTION:
     AGE = FROM 20 TO 30 <L>;
     DIZZINESS = PRESENT
        [TYPE = VERTIGO;
         COURSE = ACUTE AND PERSISTENT,
         EPISODIC [EPISODE DURATION = MINUTES <L>, HOURS <H>;
                   OCCURRENCE = POSITIONAL <H>, ORTHOSTATIC <M>,
                          NON-SPECIFIC <L>]   ];
     HEAD PAIN = PRESENT <L> [PREDOMINANT LOCATION = PERIAURAL];
     NEUROLOGIC SYMPTOMS = HEARING LOSS BY HISTORY <H>, TINNITUS <H>;
     PULSE DURING DIZZINESS = MARKED TACHYCARDIA <L>;
     NEUROLOGIC SIGNS = NYSTAGMUS [TYPE = HORIZONTAL, ROTATORY],
                        IMPAIRED HEARING <H>                          ]
```

FIG. 3. The knowledge base of System D currently consists of a set of data structures called DESCRIPTIONs such as that shown here for Meniere's Disease.

to know about three conventions being used: symbolic probabilities, separation of causal and non-causal associations, and elaboration.

Symbolic probabilities, indicated in angular brackets in Fig. 3, are subjective, non-numeric estimates of how frequently an event occurs. While exact probabilities of diagnostic associations are usually not available, a great deal of descriptive information about diagnosis exists in the form

x *frequently* causes y,

x *can* cause y,

x is *never* associated with y,

x is *commonly* associated with y,

x is *rare* (*common, very common, . . .*), and

x *only* occurs if y,

where x is some disorder and y is some fact about a case. Symbolic probabilities capture this coarse but useful information. The five[†] possible estimates we use are:

A = always,

H = high likelihood,

M = medium likelihood,

L = low likelihood, and

N = never.

Thus, the "⟨L⟩" following MENIERE'S DISEASE in Fig. 3 indicates that this disorder is relatively uncommon, and the "⟨H⟩" on the last line of the DESCRIPTION indicates that Meniere's disease often causes impaired hearing.

The second convention used in DESCRIPTIONs is the separation of causal and non-causal associations. Certain features of a disorder can be viewed as being caused by the disorder being described. For example, in medicine loss of vision, chest pain, dizziness and confusion are all abnormalities that conceptually are caused by some underlying problem. We have been using the term *manifestations* for these causally-related features. In contrast, other features of a disorder are not causally associated with it. For example, a patient's age and sex may provide very significant information about the likelihood of a certain disease being present, but they are not caused by

---

† Five is obviously somewhat arbitrary, but it has proven sufficient for our applications so far.

that disease. Features such as these will be referred to as *setting factors*. Which features in a knowledge base are manifestations and which are setting factors are indicated in the database schema specified by the knowledge base author (see Reggia, 1981). In the DESCRIPTION in Fig. 3, only the first assertion concerning age specifies a setting factor, while each of the other assertions specify manifestations.

Finally, elaboration provides further details about a manifestation and is indicated as part of an assertion inside of square brackets. For example, in Fig. 3 the assertion

```
DIZZINESS=PRESENT
[TYPE=VERTIGO;
 COURSE=ACUTE AND PERSISTENT,
          EPISODIC . . .]
```

elaborates on the type of dizziness manifested by Meniere's Disease by indicating that it is vertiginous in nature and that it occurs either in an acute, persistent fashion or in episodes.

With the above conventions in mind, the DESCRIPTION in Fig. 3 should now be relatively understandable. It indicates that Meniere's disease is a relatively uncommon cause of dizziness (because of "⟨L⟩" immediately following the name of the disease). The dizziness it causes is vertiginous in nature and either acute and persistent or episodic. When episodic, the episodes usually last for hours and are especially produced by positional changes of the head. Meniere's Disease occasionally causes periaural headache, frequently causes hearing loss and tinnitus (ringing in the ears), and so forth.

What is most important here in the context of the set covering model is that the DESCRIPTION associated with any disorder d_i specifies, among other things, the set man(d_i) of all manifestations caused by d_i. Thus, a knowledge base containing a set of disorders along with all of their DESCRIPTIONs completely specifies the information needed to solve diagnostic problems as they were defined earlier. Returning to our example, the knowledge base of System D consists of a listing of 50 causes of dizziness and their DESCRIPTIONs, each similar to that illustrated in Fig. 3. The knowledge base thus explicitly specifies the set **D** of all causative disorders of dizziness as well as the set man(d_i) for each disorder. Furthermore, the set **M** is implicitly specified, as it consists of every manifestation listed in any of the DESCRIPTIONs. The relationship **C** and the sets causes(m_j) are also implicitly specified by the collective information in the DESCRIPTIONs. As explained earlier, the DESCRIPTIONs in this knowledge base contain additional information about setting factors and estimates of relevant probabilities.

THE INFERENCE MECHANISM: A SEQUENTIAL HYPOTHESIZE-AND-TEST PROCESS

In adapting the set covering model for use in a real-world expert system several issues must be addressed and resolved. Perhaps the most obvious of these issues is the fact that diagnostic problem-solving is inherently sequential in nature. Rather than knowing all of the manifestations which are present in a specific case at the start, the human diagnostician usually begins knowing that one or a few manifestations are present, and must actively seek further information about others. In medicine, for example, the physician is typically confronted with a patient complaining of some symptom (the "chief complaint"), and must uncover other manifestations through questions, examining the patient, and laboratory-testing.

Empirical studies done over the last decade have provided convincing evidence that this sequential diagnostic reasoning is guided by a hypothesize-and-test process (Elstein, Schulman & Sprafka, 1978; Kassirer & Gorry, 1978; see Reggia, 1982, for a review). Given a few initial manifestations, the human diagnostician constructs a tentative hypothesis about the cause of those manifestations. Further information is then sought for generally two reasons: either for completeness (so-called "protocol-driven" questions), or to uncover facts specifically needed to modify the evolving hypothesis (so-called "hypothesis-driven" questions). These latter questions "test" the validity of the hypothesis, possibly confirming or eliminating part of it.

This sequential diagnostic process can be captured in terms of the set covering model presented earlier. The tentative hypothesis at any point during problem-solving is defined to be the solution for those manifestations already known to be present, assuming, perhaps falsely, that no additional manifestations will be subsequently discovered. To construct and maintain a tentative hypothesis like this, three simple data structures prove useful:

MANIFS: the set of manifestations known to be present so far;
SCOPE: causes(MANIFS), the set of all disorders d_i for which at least one manifestation is already known to be present; and
FOCUS: the tentative solution for just those manifestations already in MANIFS; FOCUS is presented as a collection of generators.

The term "generator" used here needs further definition. Rather than representing the solution to a diagnostic problem as an explicit list of all possible explanations for M^+ or MANIFS, it is advantageous to represent the disorders involved as a collection of explanation generators. An explanation *generator* is a collection of sets of "competing" disorders that implicitly represent a set of explanations in the solution and can be used to generate them. A generator is analogous to a Cartesian set product, the difference being that the generator produces unordered sets rather than ordered tuples. To illustrate this idea, consider the example diagnostic problem presented earlier (Table 1). Two generators are sufficient to represent the solution to that problem: $\{d_1\ d_2\} \times \{d_7\ d_8\ d_9\}$ and $\{d_3\ d_4\} \times \{d_8\}$. The second generator here implicitly represents the two explanations $\{d_3\ d_8\}$ and $\{d_4\ d_8\}$, while the first generator represents the other six explanations in the solution.

There are at least three advantages to representing the solution to a diagnostic problem as a set of generators. First, this is usually a more compact form of the explanations present in the solution. Second, generators are a very convenient representation for developing algorithms to process explanations sequentially (see below). Finally, and perhaps most important, generators are closer to the way the human diagnostician organizes the possibilities during problem solving (i.e. the "differential diagnosis").

Using the three data structures MANIFS, SCOPE and FOCUS, a hypothesize-and-test algorithm based on the set covering model can perform diagnostic problem solving. The FOCUS represents the tentative or working hypothesis at any point during problem-solving. The algorithm, described informally, is:

(1) Get the next manifestation m_j.
(2) Retrieve causes(m_j) from the knowledge base.

(3) MANIFS ← MANIFS ∪ {m_j}.
(4) SCOPE ← SCOPE ∪ causes(m_j).
(5) Adjust FOCUS to accommodate m_j.
(6) Repeat this process until no further manifestations remain.

Thus, as each manifestation m_j that is present is discovered, MANIFS is updated simply by adding m_j to it. SCOPE is augmented to include any possible causes d_i of m_j which are not already contained in it (derived by taking the union of causes(m_j) and SCOPE). Finally, FOCUS is adjusted to accommodate m_j based on intersecting causes(m_j) with the sets of diseases in the existing generators. These latter operations are done such that any explanation which can no longer account for the augmented MANIFS (which now includes m_j) are eliminated.

The key step in this process is Step 5, the adjustment of the FOCUS or working hypothesis. Perhaps the best way to understand this step is to follow a simple example. Recall the abstract knowledge base illustrated in Table 1, and consider the same diagnostic problem $M^+ = \{m_1\ m_4\ m_5\}$ that was used earlier. The order in which information about manifestations is discovered is determined by question generation heuristics, as described later in section 5. For now, suppose that the sequence of events occurring during problem-solving were ordered as listed in Fig. 4. What happens during problem-solving is as follows.

Events in order of their discovery	MANIFS	SCOPE	FOCUS
Initially	∅	∅	∅
m_1 present	$\{m_1\}$	$\{d_1\ d_2\ d_3\ d_4\}$	$\{d_1\ d_2\ d_3\ d_4\}$
m_2 absent	$\{m_1\}$	$\{d_1\ d_2\ d_3\ d_4\}$	$\{d_1\ d_2\ d_3\ d_4\}$
m_3 absent	$\{m_1\}$	$\{d_1\ d_2\ d_3\ d_4\}$	$\{d_1\ d_2\ d_3\ d_4\}$
m_4 present	$\{m_1\ m_4\}$	$\{d_1\ d_2\ d_3\ d_4\ d_5\ d_8\}$	$\{d_1\ d_2\}$
m_5 present	$\{m_1\ m_4\ m_5\}$	$\{d_1\ d_2\ d_3\ d_4\ d_5\ d_7\ d_8\ d_9\}$	$\{d_1\ d_2\} \times \{d_7\ d_8\ d_9\}$ *and* $\{d_8\} \times \{d_3\ d_4\}$
m_6 absent	$\{m_1\ m_4\ m_5\}$	$\{d_1\ d_2\ d_3\ d_4\ d_5\ d_7\ d_8\ d_9\}$	$\{d_1\ d_2\} \times \{d_7\ d_8\ d_9\}$ *and* $\{d_8\} \times \{d_3\ d_4\}$

FIG. 4. Sequential problem-solving using the set covering model.

Initially, MANIFS, SCOPE and FOCUS are all empty. When m_1 is discovered to be present, m_1 is added to MANIFS, and the new SCOPE is the union of the old SCOPE with causes(m_1). Since previously there were no generators in the FOCUS, the intersection of causes(m_1) with them is trivially empty. In such situations a new generator is created, in this case consisting of causes(m_1). In the terms defined earlier, this generator represents a solution for $M^+ = \{m_1\}$. It tentatively postulates that there are four possible explanations for M^+, any one of which consists of a single disease. The FOCUS thus asserts that "d_1 or d_2 or d_3 or d_4 is present".

The absence of m_2 and m_3 do not change this initial hypothesis. However, when m_4 is discovered to be present, MANIFS and SCOPE are augmented appropriately.

A new FOCUS is developed, representing the intersection of causes(m_4) with the single set in the only pre-existing generator set in FOCUS. Note that the new generator $\{d_1 d_2\}$ in the FOCUS that results from this intersection operation represents precisely all explanations for the augmented MANIFS. This new FOCUS also illustrates another important point. As information about each possible manifestation becomes available, the FOCUS changes incrementally with a monotonic decrease in the number of explanations it represents (with the exception of situations where the FOCUS becomes empty).

When m_5 is noted to be present, MANIFS and SCOPE are again adjusted appropriately. However, in this case the intersection of causes(m_5) with the single generator set in the FOCUS is empty (none of the previous explanations represented by the old FOCUS can now cover all known manifestations). The occurrence of an empty FOCUS like this again triggers a restructuring of the FOCUS: a procedure is called that produces a new set of generators from the now augmented MANIFS and SCOPE. These new generators are based on the fact that the cardinality of any explanation now contained in the FOCUS must be exactly one greater than the cardinality of its old explanations. Thus, when m_5 is found to be present, the new generators represent explanations consisting of two diseases.

Since m_6 is absent, the final solution to the problem is given by these same two generators (last line in Fig. 4). Note that these two generators implicitly represent the eight explanations for M^+ that were listed earlier. It is also interesting to note that d_3 and d_4, eliminated from the FOCUS when m_4 was found to be present, are once again viable possibilities. Had this been a larger knowledge base with additional manifestations, the FOCUS would have continued to evolve using similar set intersection operations.

4. An example application of the model

In the interest of clarity, we have so far ignored several aspects of real-world diagnostic problem-solving in presenting the set cover model. Rather than immediately examining these details, we will first briefly illustrate an interactive session with System D.

Expert System D must handle situations where multiple causes of dizziness are present simultaneously. The following conversation illustrates just such a situation. User typing is underlined and ellipsis indicates where material has been excised for brevity. The system's tracing mechanism has been turned on producing the normally unseen output highlighted in boxes which represents periodic "snapshots" of the FOCUS.

.

.

.

```
DIZZINESS DIAGNOSIS DECISION SUPPORT SYSTEM
TYPE 'CONTINUE' TO BEGIN

READY FOR COMMAND:
assert dizziness.
```

(Rather than waiting to be asked questions, the user elects to volunteer some initial information.)

FOR DIZZINESS=PRESENT

TYPE:
  (1) DEFINITE ROTATIONAL SENSATION
  (2) SENSATION OF IMPENDING FAINT
  (3) ILL-DEFINED LIGHT-HEADEDNESS
  (4) LOSS OF BALANCE WITHOUT HEAD SENSATION

=?2/3.
COURSE:
  (1) ACUTE OR SUBACUTE PERSISTENT DEFICIT LASTING MORE THAN ONE
      HOUR
  (2) ONE OR MORE CLEAR-CUT EPISODES
  (3) GRADUAL ONSET OR SLOW PROGRESSION WITHOUT CLEAR-CUT EPISODES
=?2.

(The user indicates that the patient has episodic dizziness described as faintness or light-headedness. The answer "2/3" means "2 *or* 3".)

.

.

.

READY FOR COMMAND:
assert blood pressure examination=significant
  orthostatic hypotension.

(The user indicates that the patient's blood pressure falls when he stands up, i.e. there is orthostatic hypotension.)

READY FOR COMMAND:
obtain diagnosis.

(The user activates System D's inference mechanism, which is based on the set covering process explained above.)

```
FOCUS:
 GENERATOR:
  MULTIPLE SYSTEM ATROPHY
  IDIOPATHIC ORTHOSTATIC HYPOTENSION
  SYMPATHOTONIC ORTHOSTATIC HYPOTENSION
  AUTONOMIC NEUROPATHY
  ORTHOSTATIC HYPOTENSION SECONDARY TO PHENOTHIAZINES
  ORTHOSTATIC HYPOTENSION SECONDARY TO ANTIDEPRESSANTS
  ORTHOSTATIC HYPOTENSION SECONDARY TO L-DOPA
  ORTHOSTATIC HYPOTENSION SECONDARY TO ANTIHYPERTENSIVE
    MEDICATIONS
  ORTHOSTATIC HYPOTENSION SECONDARY TO DIURETICS
  HYPERBRADYKINISM
  ORTHOSTATIC HYPOTENSION SECONDARY TO PARKINSONISM
  ORTHOSTATIC HYPOTENSION SECONDARY TO PREVIOUS SYMPATHECTOMY
  ORTHOSTATIC HYPOTENSION SECONDARY TO PROLONGED RECUMBENCY
```

(The initial FOCUS for the two known manifestations, dizziness and orthostatic hypotension, consists of a single generator which in turn consists of a single set of competing diseases. Each disease represents a minimal cover for the two manifestations.)

NEURO-OTOLOGICAL SYMPTOMS ASSOCIATED WITH DIZZINESS:
  (1) DIPLOPIA
  (2) LOSS OF OR BLURRED VISION
  (3) SCINTILLATING SCOTOMAS
  (4) FOCAL SYMPTOMS REFERRABLE TO CNS
  (5) SYNCOPE
  (6) HEARING LOSS BY HISTORY
  (7) TINNITUS
=?3&5.

(Based on the active hypothesis, System D generates its first hypothesis-driven question and discovers two new manifestations, syncope and scintillating scotomas.)

```
FOCUS:
  GENERATOR:
   BASILAR MIGRAINE
        &
   ORTHOSTATIC HYPOTENSION SECONDARY TO PROLONGED RECUMBENCY
   ORTHOSTATIC HYPOTENSION SECONDARY TO SYMPATHECTOMY
   ORTHOSTATIC HYPOTENSION SECONDARY TO PARKINSONISM
   HYPERBRADYKINISM
   ORTHOSTATIC HYPOTENSION SECONDARY TO DIURETICS
   ORTHOSTATIC HYPOTENSION SECONDARY TO ANTIHYPERTENSIVE
     MEDICATIONS
   ORTHOSTATIC HYPOTENSION SECONDARY TO L-DOPA
   ORTHOSTATIC HYPOTENSION SECONDARY TO ANTIDEPRESSANTS
   ORTHOSTATIC HYPOTENSION SECONDARY TO PHENOTHIAZINES
   AUTONOMIC NEUROPATHY
   SYMPATHOTONIC ORTHOSTATIC HYPOTENSION
   IDIOPATHIC ORTHOSTATIC HYPOTENSION
   MULTIPLE ATROPHY
```

(None of the previous possible explanations of the patient's manifestations can now account for the new manifestations. In particular, SCINTILLATING SCOTOMAS is not explained. A new FOCUS is therefore created consisting of a single generator representing minimal covers containing two disorders. The "&" here represents the "×" used in generators as described in section 3.)

CURRENT MEDICATIONS:
  (1) ANTICOAGULANTS
  (2) LARGE AMOUNTS OF QUININE
  (3) PHENOTHIAZINES
  (4) ANTIHYPERTENSIVE AGENTS

```
. . .
(12) BARBITURATES
=?2&3&4.
```

```
ELEMENTARY DISORDERS NOW CATEGORICALLY REJECTED:
DIZZINESS SECONDARY TO BARBITURATES
OTOTOXICITY SECONDARY TO AMNIOGLYCOSIDES
OTOTOXICITY SECONDARY TO SALICYLATES
ORTHOSTATIC HYPOTENSION SECONDARY TO ANTIDEPRESSANTS
ORTHOSTATIC HYPOTENSION SECONDARY TO L-DOPA
ORTHOSTATIC HYPOTENSION SECONDARY TO DIURETICS
```

(After learning the patient's medications, System D is able to narrow down the number of possibilities. How this is done is described in the next section.)

```
ABNORMALITIES ON NEUROLOGICAL EXAMINATION:
(1) OPTIC ATROPHY
(2) PAPILLEDEMA
(3) HOMONYMOUS FIELD CUT
(4) NYSTAGMUS
(5) IMPAIRED HEARING
(6) PERIPHERAL NEUROPATHY
(7) PARKINSONISM
(8) FOCAL CNS FINDINGS
=?3&4&5.
```

(The user indicates the presence of three additional manifestations: homonymous hemianopsia, nystagmus, and impaired hearing.)

```
FOCUS:
 GENERATOR:
  BASILAR MIGRAINE
      &
  COGAN'S SYNDROME
  OTOTOXICITY SECONDARY TO QUININE
  OTOSCLEROSIS
  LABYRINTHINE FISTULA
  MENIERE'S DISEASE
      &
  ORTHOSTATIC HYPOTENSION SECONDARY TO PROLONGED RECUMBENCY
  ORTHOSTATIC HYPOTENSION SECONDARY TO PREVIOUS SYMPATHECTOMY
  ORTHOSTATIC HYPOTENSION SECONDARY TO PARKINSONISM
  HYPERBRADYKINISM
  ORTHOSTATIC HYPOTENSION SECONDARY TO ANTIHYPERTENSIVE
     MEDICATIONS
  ORTHOSTATIC HYPOTENSION SECONDARY TO PHENOTHIAZINES
  AUTONOMIC NEUROPATHY
  IDIOPATHIC ORTHOSTATIC HYPOTENSION
  MULTIPLE SYSTEM ATROPHY
```

(The new FOCUS resulting from the additional information is illustrated here. Each explanation now consists of three disorders. The previous explanations could not account for the impaired hearing or nystagmus, and these new manifestations are now assumed to be due to one of five causes: COGAN'S SYNDROME, etc. Note that this FOCUS represents $1 * 5 * 9 = 45$ potential explanations in a compact fashion.)

READY FOR COMMAND:
display value (diagnosis).

  BASILAR MIGRAINE <A>
    &
  OTOTOXICITY SECONDARY TO QUININE <H>
  OTOSCLEROSIS <M>
  LABYRINTHINE FISTULA <L>
  MENIERE'S DISEASE <L>
    &
  ORTHOSTATIC HYPOTENSION SECONDARY TO ANTIHYPERTENSIVE
    MEDICATIONS <H>
  ORTHOSTATIC HYPOTENSION SECONDARY TO PHENOTHIAZINES <H>
  IDIOPATHIC ORTHOSTATIC HYPOTENSION <M>
  AUTONOMIC NEUROPATHY <M>
  MULTIPLE SYSTEM ATROPHY <L>
  ORTHOSTATIC HYPOTENSION SECONDARY TO PARKINSONISM <L>

READY FOR COMMAND:

This final diagnosis offered by System D, including a ranking of competing alternatives which will be explained below, means: "The patient has basilar migraine. In addition, the patient also probably has ototoxicity secondary to the quinine he is taking, although he could have otosclerosis or even one of the other unlikely inner ear disorders listed. Finally, the patient also has orthostatic hypotension which is probably due to his medications, but might be due to one of the other listed causes". This final diagnostic account of the patient's complex set of signs and symptoms is very plausible.

5. From model to functioning expert system

As noted earlier, the implementation of functioning expert systems like System D based on the set covering model requires that several issues be addressed and resolved. We have already discussed adopting the model to sequential problem solving, so we now turn to several other aspects of real-world diagnostic problem-solving. Further details about these issues can be found in the references (Reggia, 1981).

QUESTION GENERATION AND TERMINATION CRITERIA

The vast majority of questions generated by Expert System D, representing Step 1 in the informal sequential algorithm presented earlier, fall into the category of hypothesis-driven questions. In other words, each question is based solely on the disorders in the FOCUS at that point during problem-solving. Let us say that a disorder is *active* if it is currently in the FOCUS. Then to select its next question the expert system extracts from the DESCRIPTION of each active disorder the first attribute in an assertion whose current value is not yet known (recall that assertions, attributes and values were defined in section 3, "Case-Specific Database"). From these candidate attributes, the one in the largest number of DESCRIPTIONs of active disorders is selected to form the basis of the next question.

This simple, heuristic approach to question generation makes no claim to optimality. However, it does have certain properties that make it a useful strategy to follow. Since it selects one of the most commonly referred to attributes of active disorders, it usually produces questions that help to discriminate among the competing explanations in the FOCUS. In addition, since it selects candidate questions from the *first* unknown attributes remaining in these DESCRIPTIONs, it allows the knowledge base author to exert partial control over the order in which questions are generated (i.e. by consistently ordering the assertions in DESCRIPTIONs in a similar fashion). Finally, this approach to question generation has the advantage of being computationally inexpensive when compared with more elaborate optimization schemes that might be used.

Once a new question has been asked and answered by the user, another hypothesize-and-test cycle begins. This continues until no further questions can be generated because no assertions in the DESCRIPTIONs of active disorders contain attributes whose values have not been acquired from the user. This termination condition is a somewhat arbitrary approach to deciding when sufficient information is known. While it asks about all attributes relevant to ranking the competing explanations involved at termination time, it might leave some information unsought. To permit the knowledge base author to insure the level of completeness of information collection that is desired from an expert system, protocol-driven questions that should always be asked may optionally be included as explicit instructions to an expert system at the time it is constructed.

SETTING FACTORS AND THE RANKING OF COMPETING DISORDERS

Once the termination condition is satisfied, expert systems like System D enter a final scoring phase during which competing disorders are ranked relative to one another for the first time. In other words, the hypothesize-and-test control cycles have previously only been concerned with the construction of all possible explanations (a differential diagnosis) without regard to their relative likelihood. For each active disorder at termination time two numeric scores are calculated: a *setting score* and a *match score*. These scores are calculated using the symbolic probabilities in the knowledge base as well as any symbolic probabilities incorporated in a user's response to questions. A simple weighting scheme $(A = 4, H = 3, \ldots, N = 0)$ is used for these calculations.

The setting score for an active disorder is initialized to the numerical equivalent of its symbolic probability originally specified following its name in the knowledge base. This initial score is then incrementally adjusted upwards or downwards based only on assertions about setting factors in its DESCRIPTION. The setting score is intended to provide a generalization of the concept of prior probability in that it reflects the general likelihood of a disorder in the context of the specific setting in which it is occurring.

The match score of an active disorder is based only on M^+ and the assertions about manifestations in its DESCRIPTION. The match score is also derived using a simple weighting scheme. At termination time, an expert system conceptually has in the FOCUS all of the possible competing explanations for M^+. It can therefore derive a match score for any disorder based on the "best" explanation which contains that disorder (i.e. the explanation that as a whole would be most likely to cause M^+). For example, if d_1 is in two explanations, then the match score for d_1 would be based on its role in the "better" of these two possible explanations. Furthermore, a manifestation can be assigned to the disorder in an explanation which is most likely to be producing it in situations where that manifestation can be caused by more than one of the disorders in the explanation. The match score is intended as a measure of how closely a disorder fits the manifestations of a case, irrespective of the setting in which they are occurring.

A final score is calculated for each active disorder based on both its setting score and its match score. Since this final numerical weighting is intended to provide only a "ballpark" indication of how likely the disorder is, it is subsequently converted back into a symbolic probability to emphasize its imprecise and heuristic nature. This was illustrated in the conversation with System D when the final diagnostic possibilities were listed in order of likelihood.

It should be appreciated that the set covering model used in this fashion permits scoring which can be considered to be truly context-dependent. Not only does a disorder's likelihood depend on the specific environment in which it is occurring (setting score), but it also depends on what other disorders are postulated to be simultaneously present in an explanation, and on which of several competing explanations contain it (match score).

One final point about the use of symbolic probabilities needs to be made. While the ranking of competing disorders is done *after* the termination condition is satisfied, the symbolic probabilities A and N are used in one other way *during* the hypothesize-and-test control cycles. They are used to determine when any disorder d_i should be *categorically rejected* by the inference mechanism. For example, the DESCRIPTION of ORTHOSTATIC HYPOTENSION SECONDARY TO L-DOPA in System D's knowledge base contains the categorical assertion

CURRENT MEDICATIONS=L-DOPA <A>.

Thus, if System D discovered that a patient was not taking L-DOPA, ORTHOSTATIC HYPOTENSION SECONDARY TO L-DOPA would be immediately discarded from any further consideration by the inference mechanism (as occurred after the question on CURRENT MEDICATIONS in the conversation with System D earlier). In effect, what occurs is that the set D is changed: the set of all possible disorders is modified by removing any disorders discovered to be categorically rejected during problem-solving. All subsequent development of the SCOPE and FOCUS by the inference mechanism reflects this change in the very framework of the problem.

PROBLEM DECOMPOSITION

Since finding a minimal set cover is known to be NP complete (Karp, 1972), the task of constructing the solution to a diagnostic problem is potentially combinatorally expensive as the size of an explanation increases. This difficulty is only academic for some classes of diagnostic problems. For example, it is not uncommon for a patient seen by a physician to have more than one disease simultaneously, but it would be exceedingly rare for someone to have more than 50 diseases simultaneously. However, since the potential for combinatorial explosion exists, it is still important to address the question of when a diagnostic problem can be reduced or decomposed into smaller, independent subproblems.

One example of when this can be done is best presented by introducing the concept of "connected" manifestations. Two manifestations m_a and m_b are said to be *connected* if either causes(m_a) and causes(m_b) have a non-empty intersection, or there exists a finite set of manifestations $\{m_1, m_2, \ldots, m_n\}$ such that $m_1 = m_a$, $m_n = m_b$, and each m_j is connected to m_{j+1}. All of the manifestations appearing in Table 1, for example, are connected to one another. It can be shown that if M^+ can be partitioned into N subsets of connected manifestations, each subset of which contains no manifestation connected to another manifestation in a different subset, then the original diagnostic problem can be partitioned into N independent subproblems. The generators for the solution to the original problem are then easily constructed by appending in an appropriate fashion the generators for the solutions to the subproblems (Reggia, 1981).

Furthermore, sequentially constructing and maintaining independent subproblems in this way, each with its own SCOPE, FOCUS and MANIFS, is relatively easy. When a new manifestation m_i is found to be present, the set causes(m_i) is intersected with the SCOPE of each pre-existing subproblem. When this intersection is non-empty, m_i is said to be *related* to the corresponding subproblem. There are three possible results of identifying the subproblems to which m_i is related. First, m_i may not be related to any pre-existing subproblems. In this case, a new subproblem is created, with MANIFS $= \{m_i\}$, SCOPE $=$ causes(m_i), and FOCUS $=$ a single generator consisting of the single set of competing disorders found in causes(m_i). This is what always occurs when the first manifestation becomes known, as was illustrated in Fig. 4. Second, m_i may be related to exactly one subproblem, in which case m_i is assimilated into that subproblem as described earlier and illustrated with m_4 and m_5 in Fig. 4. Finally, m_i may be related to multiple existing subproblems. In this situation, these subproblems are "joined" together to form a new subproblem, and m_i is then assimilated into this new subproblem (not illustrated in Fig. 4 nor in the conversation with System D, both of which involved only a single subproblem).

OTHER CONSIDERATIONS

Many other considerations go into expanding the generality and robustness of the set covering model for use in real world expert systems. We will mention just three of these issues here: unexplainable manifestations, the single-disorder constraint, and non-independent disorders.

Assuming that an expert system's knowledge base and a relevant case are both correct and complete, the set cover model as described above can handle a broad range of diagnostic problems. Unfortunately, in the real world, this ideal situation is sometimes not present. A knowledge base might be incomplete or contain errors, especially during system development, and a user might enter incorrect information about a problem.

One example of such a situation is the *unexplainable manifestation*: a manifestation m_j whose associated set causes(m_j) is empty. If undetected, such a manifestation would result in repeated futile attempts by the inference mechanism to create progressively larger and larger explanations to account for all known manifestations. The important point here is that the inference mechanism must continuously monitor for unexplainable manifestations at run time. This is because an initially non-empty set causes(m_j) could potentially become empty during problem-solving if all of the disorders in it were discovered to be categorically rejected. Our expert systems currently handle this anomaly by informing the user of the situation, discarding the unexplainable manifestation, and offering the user the option of continuing with the understanding that all is not well.

Another issue deserving special attention is situations where only one disorder is expected to occur at a time. Even though such a *single-disorder constraint* may not be strictly correct in a theoretical sense, there are situations where such an assumption is justified by practical considerations. An example of an expert system called System P which uses the set covering model with the single disorder constraint is given in Appendix B. System P uses this constraint because of the exceedingly low likelihood that two of the individually very rare disorders in its knowledge base would occur in a single individual. The advantage of using the single-disorder constraint when appropriate is that it permits the automatic recognition by the inference mechanism of potential errors. This is illustrated in the conversation with System P in Appendix B when it indicates that no single disorder can account for all of the facts in the case under consideration. Such a situation might have been due to (i) user error in describing the case, (ii) an incomplete or incorrect knowledge base, or (iii) a patient with a previously unknown form of peroneal muscular atrophy.

When the single-disorder constraint is employed, two adjustments are made to the inference mechanism of expert systems using the set covering model. First, at the start of a case the FOCUS is initialized to a single generator whose single set includes all possible disorders within the domain of the expert system. This initial FOCUS represents the initial hypothesis that exactly one possible disorder is present. Second, the inference mechanism as usual monitors for the occurrence of an empty FOCUS, but interprets such an occurrence as an anomaly. It does not try to construct explanations containing two disorders, but indicates to the user that it cannot explain the current case findings with a single disorder (see conversation with System P).

Finally, we have assumed so far that the disorders in D are independent of one another, an assumption that may not be valid in some domains. One possible approach to this non-independence would be to award a "bonus" during scoring to explanations where associated disorders were involved [this was the approach used in INTERNIST; see Pople, Myers & Miller (1975)]. We have elected to study instead those situations where disorders can be partitioned into classes, with disorders in one class causing disorders in another. For example, one expert system currently being constructed involves both localization of damage in the nervous system and diagnosis.

6. Discussion

This paper has proposed the construction and maintenance of minimal set covers ("explanations") as a general model of diagnostic reasoning and has illustrated its use as an inference method for diagnostic expert systems. The set cover model is attractive in that it directly handles multiple simultaneous disorders, it can be formalized, it is intuitively plausible, and it is justifiable in terms of past empirical studies of diagnostic reasoning (e.g. Elstein *et al.*, 1978; Kassirer & Gorry, 1978). To our knowledge the analogy between the classic set covering problem and general diagnostic reasoning has not previously been examined in detail, although some related work has been done [for example, assignment of HLA specificities to antisera; Nau, Markowsky, Woodbury & Amos (1978) and Woodbury, Ciftan & Amos (1979)].

The set cover model provides a useful context in which to view past work on diagnostic expert systems. In contrast to the set cover model, most diagnostic expert systems that use hypothesize-and-test inference mechanisms or which might reasonably be considered as models of human diagnostic reasoning depend heavily upon the use of production rules (e.g. Aikins, 1979; Mittal, Chandrasekaran & Smith, 1979; Pauker, Gorry, Kassirer & Schwartz, 1976). These systems use a hypothesis-driven approach to guide the invocation of rules which in turn modify the hypothesis. A rule-based hypothesize-and-test process does not provide a convincing model of what has been learned about human diagnostic reasoning in the empirical studies cited earlier. Furthermore, rules have long been criticized as a representation of diagnostic knowledge (e.g. Reggia, 1978), and their invocation to make deductions or perform actions does not capture in a general sense such intuitively attractive concepts as coverage, minimality, or explanation.

Perhaps the previous diagnostic expert system whose inference method is closest to the set cover model is INTERNIST (Pople *et al.*, 1975). INTERNIST is a large and well-known expert system that represents diagnostic knowledge in a DESCRIPTION-like fashion and does not rely on production rules to guide its hypothesize-and-test process. In contrast to the set cover model, however, INTERNIST's inference mechanism uses a heuristic scoring procedure to guide the construction and modification of its hypothesis. This process is essentially serial or depth-first, unlike the more parallel or breadth-first approach implied in the set cover model. In other words, INTERNIST first tries to establish one disorder and then proceeds to establish others. This roughly corresponds to constructing and completing a single generator set in the set cover model, and then later returning to construct the additional sets for the generator. The criteria used by INTERNIST to group together competing disorders (i.e. a set in a generator) is based on a simple heuristic: "Two diseases are competitors if the items not explained by one disease are a subset of the items not explained by the other; otherwise, they are alternates (and may possibly coexist in the patient)" (Miller, Pople & Myers, 1982). In the terms of our model, this corresponds to stating that d_1 and d_2 are competitors if $M^+ - \text{man}(d_1)$ contains or is contained in $M^+ - \text{man}(d_2)$. It can be proven that while this simple heuristic may generally work in constructing a differential diagnosis, there are clearly situations for which it will fail to correctly group competing disorders together.† Reportedly, the serial or depth-first approach used in INTERNIST resulted in less than optimal performance (Pople, 1977; Miller *et al.*, 1982), and it has been criticized as "*ad hoc*" by some individuals working in statistical pattern classification because of the lack of a formal

---

† For example, suppose $M^+ = \{m_1 \cdots m_8\}$ and only $d_1, d_2,$ and d_3 have been evoked where $M^+ \cap \text{man}(d_1) = \{m_2\ m_4\ m_5\ m_6\ m_7\ m_8\}$, $M^+ \cap \text{man}(d_2) = \{m_3\ m_4\ m_5\ m_6\ m_7\ m_8\}$, and $M^+ \cap \text{man}(d_3) = \{m_1\ m_2\ m_3\}$. In the set cover model, $\text{Sol}(P) = \{\{d_1\ d_3\}\ \{d_2\ d_3\}\}$ which can be represented by the single generator $\{d_1\ d_2\} \times \{d_3\}$ where d_1 and d_2 are grouped together as competitors. Suppose that d_1 was ranked highest by the INTERNIST heuristic scoring procedure. Then $M^+ - \text{man}(d_1) = \{m_1\ m_3\}$ and $M^+ - \text{man}(d_2) = \{m_1\ m_2\}$, so INTERNIST would apparently fail to group d_1 and d_2 together as competitors.

underlying model (e.g. Ben-Bessat *et al.*, 1980). It is also unclear that the INTERNIST inference mechanism is guaranteed to find all possible explanations for a set of manifestations. Recent enhancements in INTERNIST's successor CADUCEUS attempt to overcome some of these limitations through the use of "constrictors" to delineate the top-level structure of a problem (Pople, 1977). These changes are quite distinct from the approach taken in the set cover model, but do add a breadth-first component to hypothesis construction.

The set cover model presented here is still evolving both theoretically and in terms of its evaluation in practice. Work is clearly needed in at least three directions: further theoretical development of the model, assessment of its application in expert systems involving a broad range of real-world diagnostic problems, and assessment of its adequacy as a cognitive model. We intend to pursue these issues in the future.

This research was supported by NIH grants 5 K07 NS 00348 and 1 P01 NS 16332 and in part by NSF grant MCS81-17391. Computer time was provided by the Computer Science Center of the University of Maryland.

References

AIKINS, J. (1979). Prototypes and production rules: an approach to knowledge representation for hypothesis formation. *Proceedings of the Sixth International Joint Conference on Artificial Intelligence*, Tokyo, Japan, pp. 1–3.

BEN-BASSAT, M., CARLSON, R., PURI, V., DAVENPORT, M., SCHRIVER, J., LATIF, M., SMITH, R., PORTIGAL, L., LIPNICK, E. & WEIL, M. (1980). Pattern-based interactive diagnosis of multiple disorders: the MEDAS system. *IEEE Transactions on Pattern Analysis and Machine Intelligence*, **2**, 148–160.

EDWARDS, J. (1962). Covers and packings in a family of sets. *Bulletin of the American Mathematics Society*, **68**, 494–499.

ELSTEIN, A., SHULMAN, L. & SPRAFKA, S. (1978). *Medical Problem Solving—An Analysis of Clinical Reasoning*. Cambridge, Massachusetts: Harvard University Press.

KARP, R. (1972). Reducibility among combinatorial problems. In MILLER, R. & THATCHER, J., Eds, *Complexity of Computer Computations*, pp. 85–103. New York: Plenum Press.

KASSIRER, J. & GORRY, G. (1978). Clinical problem solving: a behavioral analysis. *Annals of Internal Medicine*, **89**, 245–255.

MILLER, R., POPLE, H. & MYERS, J. (1982). INTERNIST-1, An experimental computer-based diagnostic consultant for general internal medicine. *New England Journal of Medicine*, **307**, 468–476.

MITTAL, S., CHANDRASEKARAN, B. & SMITH, J. (1979). Overview of MDX: a system for medical diagnosis. *Proceedings of the Third Annual Symposium on Computer Applications in Medical Care*, pp. 34–46. Piscataway, New Jersey: IEEE Press.

NAU, D., MARKOWSKY, G., WOODBURY, M. & AMOS, D. (1978). A mathematical analysis of human leukocyte antigen serology. *Mathematical Biosciences*, **40**, 243–270.

PAUKER, S., GORRY, G., KASSIRER, J. & SCHWARTZ, W. (1976). Towards the simulation of clinical cognition. *American Journal of Medicine*, **60**, 981–996.

POPLE, H. (1977). The formation of composite hypotheses in diagnostic problem solving: an exercise in synthetic reasoning. *Proceedings of the Fifth International Joint Conference on Artificial Intelligence*, Carnegie–Mellon University, Pittsburgh, Pennsylvania, pp. 1030–1037.

POPLE, H., MYERS, J. & MILLER, R. (1975). DIALOG: a model of diagnostic logic for internal medicine. *Proceedings of the Fourth International Joint Conference on Artificial Intelligence*, pp. 848–855. Cambridge, Massachusetts: M.I.T. Artificial Intelligence Laboratory Publications.

REGGIA, J. (1978). A production rule system for neurological localization. *Proceedings of the Second Annual Symposium on Computer Applications in Medical Care*, pp. 254–260. Piscataway, New Jersey: IEEE Press.

REGGIA, J. (1981). Knowledge-based decision support systems: development through KMS. *TR-1121*, Department of Computer Science, University of Maryland, College Park.

REGGIA, J. (1982). Computer-assisted medical decision making. In SCHWARTZ, M., Ed., *Applications of Computers in Medicine*, pp. 198–213. Piscataway, New Jersey: IEEE Press.

WOODBURY, M., CIFTAN, E. & AMOS, D. (1979). HLA serum screening based on an heuristic solution of the set cover problem. *Computer Programs in Biomedicine*, **9**, 263–273.

Appendix A: Table of symbols

Symbol	Meaning
$\in$	element of
$\subseteq$	subset of
$\leq$	less than or equal to
$\varnothing$	empty set
$\forall$	for all
$\cup$	set union
$\cap$	set intersection
$\lvert \ \rvert$	set cardinality
M	universe of manifestations (effects)
D	universe of disorders (causes)
C	causative relationship

Appendix B

Expert System P addresses the problem of diagnosing the cause of wasting of the muscles of the lower legs (peroneal muscular atrophy). The causes of this problem are very rare, and include entities such as hypertrophic neuropathy of Charcot–Marie–Tooth and Refsum's Syndrome. Because of the rarity of each of these disorders, it is extremely unlikely that more than one would occur in an individual, so System P uses the set covering model with the additional constraint that only one possible cause can be present in any case. This leads to the following interesting result (user typing is underlined in the conversation, and the system's tracing mechanism is turned on producing the output highlighted in boxes which represent normally unseen periodic "snapshots" of the FOCUS):

.

.

.

```
PERONEAL MUSCULAR ATROPHY DECISION SUPPORT SYSTEM FOR
DIAGNOSTIC CLASSIFICATION. TYPE 'CONTINUE' TO BEGIN.

READY FOR COMMAND:
continue
OK—PLEASE ANSWER THE FOLLOWING . . .
```

AGE OF ONSET:
  (1) INFANCY
  (2) CHILDHOOD OR ADOLESCENCE
  (3) ADULT LIFE
=?2.

(The manner in which the system selects questions to ask was addressed in section 5.)

```
FOCUS:
GENERATOR:
    HEREDITARY SPASTIC PARAPLEGIA WITH PERONEAL MUSCULAR
      ATROPHY
    FRIEDRICH'S ATAXIA
    REFSUM'S SYNDROME
    PROGRESSIVE SPINAL MUSCULAR ATROPHY OF CHARCOT-MARIE-
      TOOTH
    NEURONAL TYPE OF CHARCOT-MARIE-TOOTH DISEASE
    NON-DOMINANT HYPERTROPHIC NEUROPATHY OF CHARCOT-MARIE-
      TOOTH
    HYPERTROPHIC NEUROPATHY OF CHARCOT-MARIE-TOOTH
```

(Knowing that exactly one cause of peroneal muscular atrophy is present, System P starts off with an initial hypothesis consisting of a single generator containing the eight possibilities. One cause is immediately discarded because it always starts in infancy, leaving the seven potential explanations listed here.)

INHERITANCE:
  (1) SPORADIC
  (2) RECESSIVE
  (3) DOMINANT
=?unknown.
SEVERITY OF MOTOR DEFICIT:
  (1) MILD
  (2) MODERATE
  (3) SEVERE
=?3.

```
FOCUS:
 GENERATOR:
  FRIEDRICH'S ATAXIA
  REFSUM'S SYNDROME
```

(The three possible causes of a severe motor deficit are FRIEDRICH'S ATAXIA, REFSUM'S SYNDROME, and HYPERTROPHIC NEUROPATHY OF DEJERINE-SOTTAS. Intersecting this set with the previous generator decreases the alternative explanations to the two listed here.)

SENSORY DEFICIT:
  (1) PRESENT
  (2) ABSENT
=?<u>1.</u>

FOR SENSORY DEFICIT=
  PRESENT . . .

  SEVERITY:
  (1) MILD
  (2) MODERATE
  (3) SEVERE
=?<u>2.</u>

(When told that the patient has a sensory deficit, System P automatically asks for relevant elaboration about its severity. This request for elaboration is the only example of a protocol-driven question in this conversation. The discovery of a sensory deficit does not change the FOCUS, shown immediately below, because this manifestation can be caused by both active disorders.)

```
FOCUS:
  GENERATOR:
   FRIEDRICH'S ATAXIA
   REFSUM'S SYNDROME
```

PES CAVUS:
  (1) PRESENT
  (2) ABSENT
=?<u>1.</u>

```
FOCUS:
  GENERATOR:
   FRIEDRICH'S ATAXIA
```

(A third manifestation, PES CAVUS or "high arches", is uncovered, and intersecting its set of causes with the FOCUS results in a single remaining possible explanation. Rather than accepting this hypothesis as the diagnosis, System P continues to generate questions to confirm it.)

OTHER NEUROLOGICAL FINDINGS:
  (1) RETINITIS PIGMENTOSA
  (2) DEAFNESS
  (3) NYSTAGMUS
  (4) DYSARTHRIA
  (5) CEREBELLAR ATAXIA
  (6) PYRAMIDAL SIGNS
=? (MULTIPLE ANSWERS PERMITTED)
<u>1.</u>

```
A SINGLE VALUE OF
TYPE OF PERONEAL MUSCULAR ATROPHY CANNOT EXPLAIN
ALL OF THE FEATURES OF THIS CASE
SHOULD PROCESSING CONTINUE USING THOSE VALUES THAT
ARE NOT CATEGORICALLY REJECTED
=?(YES/NO)
no.

TYPE OF PERONEAL MUSCULAR ATROPHY=
 UNKNOWN

READY FOR COMMAND:  .  .  .  .
```

(When System P learns that RETINITIS PIGMENTOSA is present, intersection of its causes with the generator results in an empty FOCUS. System P gives up and classifies this patient as having an unknown disease.)

What is striking here is that System P automatically detects that it does not know the diagnosis in this case. This is because of the special constraint that only a single disorder be present, imposed by the creator of System P, contradicts the definition of adequacy required of an explanation in the context of this specific case.

On the application of expert systems

ANDREW BASDEN

Decision Support Systems Group, Corporate Management Services, I.C.I. plc, P.O. Box 11, The Heath, Runcorn, Cheshire, U.K.

Expert systems have recently been arousing much interest in industry and elsewhere: it is envisaged that they will be able to solve problems in areas where computers have previously failed, or indeed, never been tried. However, although the literature in the field of expert systems contains much on their construction, on knowledge representation techniques, etc., relatively little has been devoted to discussing their application to real-life problems.

This article seeks to bring together a number of issues relevant to the application of expert systems by discussing their advantages and limitations, their roles and benefits, and the influence that real-life applications might have on the design of expert systems software. Part of the expert systems strategy of one major chemical company is outlined. Because it was in constructing one particular expert system that many of these issues became important this system is described briefly at the start of the paper and used to illustrate much of the later discussion. It is of the plausible-inference type and has application in the field of materials engineering.

The article is aimed as much at the interested end-user who has a possible application in mind as at those working in the field of expert systems.

1. Introduction

There is something of a gap in the literature on expert systems. While much has been published on the strategies involved in their construction and on techniques for representing knowledge, little has been published on how to apply them to solve industrial and other real-life problems. But with the growing interest in expert systems in industry and elsewhere, and with the benefits they promise to give, such a gap needs to be filled.

Since many individual expert systems have been described in publications and conferences which are apparently aimed at real-life problems it is perhaps surprising that this gap exists. But authors normally describe such aspects as how their system was constructed, what knowledge representation was used and the accuracy of the results obtained, while the advantages and disadvantages of expert systems when compared with other methods are seldom discussed in any detail. Only occasionally (for example, Coombs & Hughes, 1982) has there been any discussion on the roles of expert systems and the likely benefits in such roles, and there has been little discussion on how real-life requirements should influence the development of expert system software. It is not even very clear which types of expertise could be modelled with a reasonable expectation of success, given the current technology.

For successful application of expect systems on a wide scale such issues need to be tackled. Although some aspects may have been given consideration individually, they have seldom been brought together and discussed in any detail. This article seeks to do that, by discussing:

some advantages of the expert system approach;

59

DEVELOPMENTS IN EXPERT SYSTEMS
ISBN 0-12-187580-6

roles and benefits of expert systems;

some limitations of expert systems, and whether they can be applied to "wide and shallow" knowledge domains;

part of our approach to using expert systems;

some implications for the design of expert systems.

2. One expert system project

It was initially while constructing an expert system for one particular application that many of these issues were raised for the author. So, although the purpose of this article is not merely to report on one system but rather to discuss the issues more generally, it will be described here briefly to act as an illustration of a number of general points made in later sections.

The expert system to be described is of the plausible-inference type, similar in operation to PROSPECTOR (Duda, Gaschnig & Hart, 1977) but simpler, and was constructed with the twin objectives of investigating computer techniques that might be suitable in the field of materials engineering and, assuming suitability, of being the prototype of a tool for regular use. Its function is to predict the risk of stress–corrosion-cracking in stainless steels. The prototype proved very successful and is being prepared for such use by a process involving finer tuning of some parameters in the knowledge base, extension of the knowledge base to cover a wider range of materials, some small changes to the software and careful design of the man–machine interface.

Stress–corrosion-cracking (SCC) is the cracking of metals and other materials under the combined effects of a chemical corrosive attack and tensile stress. It can occur at stresses that would not normally be sufficient to crack the material, and in environments which, on their own, would have little corrosive effect. Thus its occurrence can be unpredictable. Its economic consequences are not trivial. Occurrence of SCC in chemical plant and similar installations can result in lost production, safety hazards and expensive repair procedures. For this reason plant design has traditionally been overcautious, resulting in excess capital expenditure and constraints on operability. This, of course, is not desirable, and particularly so in times of recession, so there is an incentive to reduce over-design by more accurate estimates of the risk of SCC.

In some materials, such as stainless steels in aqueous solution, the physical and chemical mechanisms are reasonably well understood, having undergone considerable study for more than 30 years, but in other materials this is not so. Even in those situations that are understood, the problem of predicting SCC in chemical or industrial plant is not easy, because, although some mechanisms are known, there is often considerable uncertainty about the nature and magnitude of influential factors. It is not always easy, for instance, even to know the precise temperature at the metal wall, particularly if scale might have formed thereon, let alone other important factors like the concentration of chloride ions. Thus, for these reasons, much judgement is needed in predicting or diagnosing SCC. Since certain types of expert systems offered techniques to handle judgemental reasoning, they seemed worth investigating in this context.

The SCC expert system (SCCES) was created to calculate the risk of various factors involved in SCC, such as crack initiation, when evidence is supplied by the user. The evidence concerns such things as temperature of the fluid and whether there are crevices in the metal surface. It consists of a knowledge base of rules and facts specific to the SCC domain, linked to commercially-available, packaged, general purpose, plausible-inference software (Paterson, 1981; SPL, 1982). The knowledge base was constructed by two corrosion specialists and the author.

It should be noted that the SCCES does not follow the normal mould for this type of plausible-inference expert system. The knowledge base of many such systems contains a representation of diagnostic or other problem-solving procedures, many of which are rules-of-thumb gleaned from experience. In contrast, the SCCES is essentially a causal model. That is it represents some of the understanding that the corrosion specialists have of the physical and chemical mechanisms involved. In some places the causal link between one fact and another is known (such as 'if the water contains temporary hardness and the temperature is high enough then there will be precipitation") and the inference links were used to approximate it. But in others [such as "if the fluid is a concentrated solution of anything then it is likely (by virtue of impurities) to contain substances which are at the limit of their solubilities"] the causal links either are not fully understood or are complex, in which case they are represented in summary form by a rule-of-thumb. For both types of link the inference mechanisms offered by the software have so far proved adequate. Because a causal model represents generalised understanding rather than specific problem-solving rules it is likely to be more versatile, and indeed it may be the use of a causal model that led to the discovery of the variety of roles for expert systems described below. Other implications of causal models are discussed later.

No precise definition of expert systems is attempted in this article, but in what follows they are loosely assumed to be computer systems that can hold human-like knowledge of (in theory) any kind and can process knowledge in a more human-like fashion than do conventional computer systems. This includes types other than the plausible-inference system used for the SCCES (such as inductive learning and predicate calculus) and unless otherwise specified the term "expert system" is used with this wider meaning. It is also used synonymously with similar terms, like "(intelligent) knowledge-based system".

3. Advantages of expert systems

A number of advantages have been claimed for expert systems, particularly of the plausible-inference type, over "conventional" computer techniques. Indeed, the author has found that expert system techniques are more readily accepted, not among practitioners of "conventional" computing, but among the end users, especially those who have perhaps viewed computers as "useful, but they can't solve my real problems, because real problems involve a lot of judgement". Some of the advantages have been found to be of particular value in the SCCES and, although well-known to some, deserve to be restated in this context since it is from these that many of the benefits of expert systems derive.

3.1. FLEXIBILITY OF EXPRESSION

Expert systems are able to embody the rules-of-thumb that practitioners tend to carry around in their heads but never write down, as well as more formal expertise. SCC expertise contains a mixture of rules-of-thumb gained from experience, and well-known physical and chemical mechanisms. Since "real life" contains both types of knowledge, this is one of the attractive features of expert systems.

Expert systems present a possible alternative to conventional computer models, especially where the relations are known but difficult to reduce to equations or where they are too complex for the purpose of the system. Oldfield & Sutton (1978), for instance, developed a mathematical model of crevice corrosion based on physical chemistry theory. But it would not be practicable in the SCCES context because, not only does it require detail which would not normally be available, such as the compositions of solutions and the geometry of the crevice, but also the full calculation is too complex for what is a relatively minor part of the SCCES. Expert systems techniques would allow some parts of the calculation to be handled in detail and others to be summarized by plausible-inference links, giving a level of detail appropriate to the use of the system.

3.2. HUMAN-LIKE PROCESSING

The information processing performed by most types of expert system is a simple, but attractive, form of reasoning. Compared with conventional computer programs, an expert system operates at a level and in terms and concepts with which the user can feel affinity—at the level of rules and facts, of items and the relationships between them, rather than at the level of program steps. Moreover, the reasoning in the knowledge base can be examined by the user, who can ask the system, for instance, to explain why it is asking a particular question. This gives what has been called a "human window" into the internal workings, and is useful where the user needs to develop trust in the new system, and also for detecting errors of logic in the knowledge base. However, while there are exceptions (for example, Stevens *et al.*, 1981), many current human windows tend to be rudimentary, and much research is still needed. Clancey (1983) gives a very useful discussion.

It seems that causal links might give a more meaningful human window on the knowledge base than do pure rules of thumb. Compare "I am asking about crevices because experience shows that they can increase the risk of SCC" (rule-of-thumb) with "I am asking about crevices because they act as cells of local activity in which the pH drops and chloride ions can be concentrated. Both these factors can make passivity more marginal, and thus increase the risk of SCC initiation" (causal links). (Current software would render it less concisely, unfortunately.) The former may be of use to naive users of the SCCES but the most likely users are those who are experienced in the materials domain but are not SCC specialists. To such people the former reasoning is already well-known while the latter would be much more useful. We have found the production of useful human windows of this sort to be a good guide in the process of constructing knowledge bases.

3.3. EASE OF EXPRESSION

The language in which the knowledge base is expressed can be closer to the sort of language used by the specialists, than are "conventional" programming languages. Expert systems accept a statement of the knowledge directly as facts and rules instead of program steps and therefore, we have found, are much more easily understood by practitioners in the domain who might not be very computer-literate. Thus the need for a "programmer" is much reduced (although not yet always to zero), with the result that the SCCES was developed in a surprisingly short time (about 6 man-weeks of effort for a knowledge base with 100 nodes and 200 links) (see also section 6).

3.4. UNCERTAINTY

Because uncertainty, and even contradictory evidence, are handled in a natural way in plausible-inference systems, they can be used in areas of incomplete knowledge and where judgement is needed, like SCC problem-solving. This is in contrast to many "conventional" computer systems, where uncertainty and contradictory evidence, so prevalent in real life, cannot even be accepted.

In short, expert systems offer a more human-like type of computing. We have become convinced that they are thereby more easily able to tackle some areas of real life where judgement is required, than are "conventional" computer techniques. The next section describes a number of roles in which expert system techniques can be expected to offer benefits. These roles and benefits stem from advantages expert systems possess over conventional computers (as above) and over manual or human techniques, as described below.

4. Roles of expert systems

When the SCCES Project started it was expected that the main users of the expert system would be "general" materials engineers, who have expertise in a wide range of domains but who would approach the appropriate specialists for advice in difficult cases. (We will use the term "practitioner" for such people, who have some experience in a domain but do not have deep specialist understanding.) The SCCES, representing a portion of the corrosion specialists' expertise, would be able to offer similar advice and thus reduce the number of calls on the specialists' time. This is the consultancy role, but our experience in the project, and elsewhere, indicated a wider variety of roles than was at first anticipated. These are described below. Our appreciation of roles has obviously been influenced by the fact that the SCCES is of the plausible-inference type of expert system, but our experience, though limited, suggests they are not restricted to that type. For instance, a PROLOG-based data analysis system is currently being used in the knowledge-refining role described below.

Many of the benefits outlined below are regarded (by corrosion specialists as well as the author) as near certainties in the case of the SCCES. While others are probably achievable, it has not been possible to assess the problems that might be met in sufficient detail. In the discussion below it is assumed that some improvements have been made to the software used, particularly in its human window facilities and although many of the comments below are made in the context of SCC, this is mainly by way of illustration and it should be clear that many can be generalized to other domains of expertise.

The classification of roles below is as seen from the viewpoint of end users rather than that of AI researchers. This means that one should not expect different AI techniques to be needed by different roles. Indeed, not only the same techniques but the same software and even the same system will sometimes be used in several roles. The roles dimension is orthogonal to the other two dimensions: of AI techniques and of what Stefik *et al.* (1982) call a "characterization of expert tasks" (namely Interpretation, Diagnosis, Monitoring, Prediction, Planning and Design).

A number of points regarding accuracy of the results produced by an expert system are made while considering the consultancy role.

4.1. CONSULTANCY

In a consultancy role an expert system is used by a non-specialist to obtain specialist advice or other forms of help in accomplishing some task. The expert system in effect amplifies his/her expertise in the given domain, and can sometimes be used in place of a specialist.

Many existing expert systems directed at specific applications have been designed for this role: CASNET, INTERNIST, MYCIN, PROSPECTOR, Heuristic DEN-DRAL and others. Barr & Feigenbaum (1982) contains a useful summary of these. Even TEIRESIAS (Davis & Lenat, 1982), designed to construct knowledge bases for another expert system, can be seen in this light since it performs some of the specialist activities of a knowledge engineer and possesses detailed knowledge of the target expert system.

As mentioned above, this was the initial role envisaged for the SCCES, and it is still seen as a major role. To us the benefits would lie in:

greater reliability (humans can forget relevant factors, especially if under stress or in a situation where time is critical);

increased consistency in the relative importance given to different factors and in dealing with probabilities;

increased accessibility (human specialists can be on holiday, on call or otherwise unavailable);

the ability to either arrive at a faster solution or try a greater number of alternatives in the time available;

and the easier duplication of expertise (copying a disc file instead of undergoing training).

These are some advantages that expert systems have over humans, and add up to reducing the load of "trivial" enquiries on existing specialists, freeing them to give more attention to the less trivial ones and to other work.

We do not think it likely that expert systems will supplant human specialists, though. For one thing, in many domains the specialists are likely to have superior expertise for some time, if not forever. But, even if expert systems do catch up [perhaps when systems with facilities for improving their knowledge ("learning") appear on the market] the human will still have the advantage of being able to take apparently extraneous factors into account (such as economic or political factors), and to recognize which factors are relevant.

Although it has often been expected that expert systems could be used by the novice (e.g. diagnose your own illnesses and prescribe your own drugs), there are indications that this will not be as easy, nor as useful, as was at first thought. On the one hand, in any specialist field not only are there phrases and jargon words of specialized meaning, but apparently ordinary words might have special meanings of which the novice might be dangerously unaware. On the other, there are a large number of factors which practitioners in the domain take for granted but which would have to be posed to the novice explicitly. Moreover, at present, in industry it is probably more cost-effective for consultancy-role expert systems to be used in aiding practitioners rather than novices because this is where a competitive advantage is more likely to be achieved.

The SCCES offers the above benefits in two activities in which the materials engineer (or other general practitioner) is involved—in the design of mechanical structures and in remedying failures [two of the expert tasks of Stefik *et al.* (1982)]. Either two versions of the system could be built, one for design and one for remedies or, as with the SCCES, one system can be used for both. This will probably include a causal model with the possible addition of some knowledge specific to the two activities.

Parameters of a proposed design or of a failed situation can be given to such a system which, in the case of the SCCES, calculates the risk of SCC. But then the user tends to ask a number of other questions, such as the following.

The risk of SCC is too high; what parameters can I usefully alter to reduce it?

Which is the single most important parameter that caused this SCC?

I was uncertain in some of the answers I gave; are any of those parameters likely to be important?

What happens if I change these parameters? (The latter verges, of course, on simulation.)

It appears that in domains like SCC the apparent plethora of possible questions can be condensed to just a few basic types, most of which can be answered by a causal model with a versatile human window.

One possible disadvantage of a causal model in this role is that it will be slower in operation than a system designed specifically for one activity. But we have not so far found this to be a problem and the advantages of greater detail and better human window outweigh it.

Accuracy of results in this type of system is not as important as at first might be thought. If the accuracy of the system is high then an expert system like the SCCES will yield a set of parameter limits, but even if the accuracy is not high, it should at least indicate to the user which parameters are more important than others, and should still be able to highlight unexpected correlations. If the risk of chloride SCC is borderline, for instance, the precise numerical value of the risk is of little importance; of more importance is its sensitivity to changes in parameters like temperature. So if an expert system is not highly accurate and therefore unable to give precise answers, it can still give guidance and thereby help the user to increase the accuracy or certainty of his own predictions or pinpoint areas where other work should be done to improve the accuracy of estimated operating conditions. Because when an expert system is first brought into regular use its accuracy may not very high, it is important that its initial uses and roles are those where this will not prevent it giving tangible benefits, otherwise it may be rejected by its users and never achieve its full potential.

[Because accuracy has not been crucial for the SCCES, the crude piecewise-linear approximations provided by the software used have so far proved entirely satisfactory. Similarly, although there is currently much argument about which are the "correct" mechanisms to use for plausible inference, it is not likely, as Feigenbaum (1977) points out, to be very important in real-life applications at this stage. What is more important is that the human window should be versatile and well-engineered.]

It is often not sufficiently appreciated just how important non-technological consider-ations are. Even if we pay close attention to knowledge representation, to methods of eliciting knowledge from specialists, to ensuring the control structures, human window and inference methods are the best available, our system could fail in application if we fail to ensure that it has been aimed at the right class of users, that it meets their perceived needs, or that it will provide tangible benefits especially in the early stages of use. Also, careful consideration must be given to how the system is introduced to its users and how it will affect their work patterns. It has been implicitly assumed so far that the expert system will be found to have inaccuracies in its knowledge base when first used which will be corrected by a continuous process of feedback. Systems like TEIRESIAS (Davis & Lenat, 1982) provide a technological aid to this process and are likely to be very important, but it is just as important that the human organisation is adapted—to ensure inaccuracies are looked for and fed back to those who maintain the knowledge bases and that the necessary maintenance is in fact carried out. We have found Lancaster Systems Methodology (Checkland, 1981) invaluable in aiding such considerations.

Although these points concerning accuracy and non-technological considerations have been made under the consultancy role, they are relevant to the other roles.

4.2. CHECKLIST

On a number of occasions, a reduced version of the consultancy role has been thought by users to be important—that of a checklist. There has been such a measure of agreement on this role among potential users in a variety of disciplines in I.C.I. that it is worth mentioning explicitly.

As mentioned earlier, one advantage of expert systems over humans is that humans can forget. The expert system, in reliably posing all relevant questions, could act as a checklist, reminding the user of all the factors to take into account, even if no other benefits accrue. An advantage over checklists on paper is that these inevitably include questions that do not really matter in a given case but do in others, while expert systems can "intelligently" select or order the questions.

4.3. TRAINING

There is much interest in using expert systems in a training role. The trainee will sit and be taught by the computer system. But in contrast to earlier, rigid approaches at computer-aided instruction, current knowledge-based systems are often mixed-initiative systems, able to lead the trainee through a set of exercises, determine his/her specific areas of weakness and tailor the exercises to suit. SOPHIE (Brown, Burton & deKleer, 1982) for example, teaches electronic troubleshooting and GUIDON (Clancey, 1979), diagnostic problem-solving.

But most such systems are aimed at initial training. Our experience with SCCES suggests that expert systems might also be useful in continually improving the expertise of experienced, but general, practitioners. General materials engineers are likely to know there is a link between crevices and SCC (see section 3.2) but not all the physical and chemical mechanisms involved. A good human window on to a knowledge base with sufficient detail can, if they use it, give them a more specialized understanding.

4.4. REFINING EXPERTISE

The fourth role is in refining the expertise of the given domain. Most specialists will freely admit to having gaps in their knowledge, and it is likely that expert systems can help to identify where these lie. The benefit arises partly when the expert system is being constructed—"Do I really understand how these parameters interact, and would it significantly improve my understanding of the real problem if I did?". The incentive to more careful codification of existing knowledge is increased as the expert system provides a means whereby the codification can be readily used and tested. The human window facility for examining the reasoning is particularly attractive here.

Additionally, there are always points on which specialists will disagree. It is thought that expert systems like the SCCES may be of assistance here. If two different versions were constructed they could be compared by running a number of test cases and, moreover, the root of any discrepancy can be investigated, again using the human window facility. But truth maintenance systems (McAllester, 1980) may be a better approach, in that they explicitly record justifications for their beliefs. Over a period of time the expert systems can accumulate the expertise of a number of specialists and thus become a repository of knowledge in the given domain.

We believe that an expert system used in this role can also be of significant benefit as a guide to research, by highlighting weaknesses in current understanding.

4.5. COMMUNICATION MEDIUM

The fifth role is as a medium of communication, an alternative to paper. Unlike conventional computer techniques, expert systems have some of the flexibility of paper, in being usable for representing a wide variety of information. But whereas on paper—in a textbook on corrosion, for instance—it can be difficult to insert and maintain many cross-references or to locate the precise information you want, with expert systems these problems are reduced. Its knowledge base is not restricted to the two dimensions of a sheet of paper and the software can be used to aid searching and processing of what can be a large collection of rules and other information. The expert system is thus seen as an active textbook—or even as an active notebook for rough notes.

In any case, the most vital knowledge often exists in the memory of specialists or in records that are not readily accessible. Few published papers or conferences in the corrosion domain deal with real experience, as distinct from academic aspects. The knowledge-base of expert systems could act as communication medium on this desired level since it can make such "hidden" knowledge explicit.

This role overlaps both those in training and knowledge refining to some extent. But it has been mentioned separately because it makes possible a number of other roles. For instance, the expert system, like a textbook, could be used merely as a reference work. Alternatively, it could be used for explaining the rationale behind decisions—and for logging both the decisions and why they were taken. The expert system could also form the medium for sharing experience between institutions. A particular attraction is that the information exchange would be in the form of rules—or generalized experience—rather than detailed information, where complications might arise regarding confidentiality. In industry, customer technical service and the capture of expertise of specialists who are about to leave or retire are two areas where such a role looks promising.

4.4. REFINING EXPERTISE

Although this role has not arisen in the context of stress–corrosion-cracking, it has arisen on a number of other occasions. If one wishes to obtain data held in databases then there is a large amount of information one needs to know:

in which database(s) the data is held;
when each database is available;
how to connect into each;
the query language of each;
what to do when things go wrong;
etc.

Such information can be very detailed and it is a common experience that if one is not making frequent use of a particular database—or any other system—then one forgets much of it.

Much of this information is most naturally expressed in rule form ("If error XYZ occurs, do so-and-so" or "If the terminal goes dead then the system may have crashed") and tends to be subject to frequent change as experience accumulates or systems are upgraded. Sometimes an element of judgement is involved, as when deciding which of several databases hold the most reliable version of the desired data. This suggests that expert systems have advantages over conventional computer systems here, being used simply as a repository of the detailed rules on how to use certain systems. The RITA system (Anderson *et al.*, 1977) was designed for this type of role.

4.7. DEMONSTRATION VEHICLE

Our last role, apparently minor when considering application to real-life systems, but nevertheless important at this early stage in the acceptance of expert systems, is that of conveying their concepts, capabilities and constraints to a public who are unaware of them. Many people are unfamiliar with the basic concepts, and even more do not have much appreciation of how they can be employed in practice, what benefits they can expect and what limitations there are in particular circumstances. Hands-on experience is important for gaining such an appreciation, both in constructing a knowledge-base and in using it. Much valuable learning can be obtained even if the knowledge base thereby constructed has only been half developed. Now that expert system software is available for a price almost within the realms of petty cash (MicroExpert: ISIS, 1982), such experience need not be hard to obtain.

5. Limitations of expert systems

So far this article has stressed the advantages of expert systems. There are, of course, limitations. Some have to do with the nature of expert systems while others arise from the cultural and computing environments and the current state of the technology. If successful application is to be made then it is important to understand what they are. But it is not known for certain what the limitations are since they have not been adequately explored, and seldom discussed.

However, our experience so far suggests a number of guidelines. As a general rule, it would be unwise to apply expert systems to:

problems which are too simple (under 10 rules, say) since the human can ordinarily handle them very adequately;

problems which are too complex (over 10,000 rules, say), since construction and search times become too long and current hardware has difficulty coping with anything larger;

problems which require none of the advantages of expert systems, such as well-structured numerical problems;

problems which rely on information more suited to processing by the human brain than by computer—see below;

problems in "wide and shallow" domains—this is discussed below.

Most of these guidelines are obvious in retrospect, but are stated here because they are not always obvious in prospect. Some are discussed in more detail by Stefik *et al.* (1982) and many will be qualified by different balances of costs and benefits in different circumstances or modified as technology advances. For instance, a five-rule expert system may indeed be worth constructing if the reliability and consistency levels that humans normally achieve are not high enough. Or more powerful hardware—or more efficient use of existing hardware—might allow systems larger than 10,000 rules.

While expert systems, like most computer systems, can process numeric or textual input, the information in many problems comes primarily through our senses of touch, smell, sound or sight and the problem has a large pattern-recognition element. For this type of problem it is probably unwise to attempt to apply expert systems at this stage since human capabilities are far superior, though as always this will depend on circumstances. While vibration monitors (with their numeric output in the form of electrical voltages) could conceivably be used, it has been said that the best way to diagnose the state of a ship's motor is through the soles of one's feet. Similarly, although it is certainly possible to construct an expert system for identifying birds from the colour of wings, bills and so on—and the author has done so—the author would not use it in the field. Even if it were pocket-sized and feather-light (which current systems certainly are not) it would take far too long to answer all the relevant questions. Far superior is the almost instant recognition achieved by even a mediocre bird-watcher from a quick glimpse of the bird—especially as he/she takes account not only of colour and shape but of habits and call as well, factors which can be difficult to express in words or numbers.

Traditionally, expert systems have been built for "deep and narrow" domains, where recognized specialists are available from whom the knowledge base can be constructed and it is known that the relevant factors lie within the fairly narrow bounds of (in the case of SCC) certain areas of physics, chemistry and engineering. Even in the types of SCC not yet well understood, it is highly unlikely that relevant factors will be found outside these bounds—the stability of the Middle East is not likely to be very relevant to SCC problem solving, for instance. So any increase in expertise in the domain is obtained by a "deeper" rather than "wider" understanding—by finer tuning and the inclusion of more special-case reasoning. "Deep and narrow" domains are often either in the realms of science or engineering, or in domains comprising man-made rules (which are complete and do not have to be discovered) such as the social security laws.

But there is discussion in expert system circles about whether they can also be used for "wide and shallow" knowledge. In these domains there is a wider range of potentially relevant factors, and there may be no specialists of the type found in the deeper, narrower domains of science and engineering. For example, in order to predict the rate of inflation, the economist must take into account a wide variety of factors and there is little agreement over which factors are important.

It might be difficult to obtain a successful expert system in such a field, for a number of reasons. First, it is difficult to distinguish relevant from irrelevant factors—the whole of life could be relevant—so accuracy and reliability would be low. Second, the expert system would be difficult to test exhaustively. Third, were such an expert system to be constructed, it would probably be impossibly slow in operation since most expert systems make much use of searching the (huge) knowledge-base.

But there are three reasons why such an argument may not be very useful. First, it is unwise to say a thing cannot be done, until someone, motivated to succeed, has tried and failed. There may be a way of overcoming some of the above problems that are not apparent at the moment. In particular, the Japanese Fifth Generation thrust (JIPDEC, 1981) is aimed at this and there are indications that they might succeed, using parallel-processing techniques.

Second, the above problems may only be insurmountable where high accuracy and complete knowledge is required from and in the expert system. In "wide and shallow" domains, human performance itself may not be very high, so an expert system merely needs to achieve better results than the human with reasonable reliability to become useful. The scales are tipped in favour of the expert system by our capacity for forgetting the vital fact. There are several systems of this type which are showing some success [such as the football pools system of George (1980) which produces a better-than-average stream of minor wins, though few major ones].

Third, as mentioned above, there are roles where the need for completeness and accuracy is reduced. The active notebook is one such role, and refinement of knowledge or expertise is another. In the latter the expert system is not so much something that can make accurate predictions as a system that allows knowledge to be expressed easily, in a form that is natural to the practitioner, and then viewed from several angles, tested, explained and compared with that of other practitioners. Since there is usually enormous scope for refinement of "wide and shallow" domains, such knowledge refinement expert systems should find many applications. The author is involved with one such project, to develop data analysis methodology.

6. Expert systems for real-life use

Thus, although there are limitations in the expert systems approach, we are convinced that real benefits can be obtained in many areas from their use. Up to now we have sought to make maximum use of commercially-available software, rather than developing our own, and it may be asked why we have taken such an approach. Also, are we likely to continue with this approach in future?

The main reason we used available packaged software at first was to speed up our gaining of applications experience and learning of concepts. Though far from perfect, plausible inference seemed usefully appropriate to many problems and seemed worth exploring in depth in applications contexts before tinkering with knowledge processing techniques. Predicate calculus is likewise being explored. Moreover, bearing in mind that I.C.I. is a chemical company, and that much research into expert system techniques is being carried out already elsewhere, there was little incentive to embark on such research of our own. If we were to carry out any research, we should not duplicate but instead investigate the many problems that lie between a working system and its regular, beneficial use in real life—an area where hardly any work had been done.

Perhaps because the simple plausible-inference mechanism seems to be widely appropriate, and useful even if not perfect (see discussion on accuracy in section 4.1), we have found a number of benefits in this approach:

construction of useful knowledge bases takes man-months rather than man-years;

paradoxically, working within the constraints of current software has led to a clearer understanding of which features are necessary for given applications than would have been obtained by working with more versatile software;

the limited knowledge representation offered by such packages is probably simpler to teach to those involved in constructing knowledge bases than that offered by more versatile AI software.

We find we are thus able to teach domain specialists some knowledge engineering skills so that they can construct their own knowledge bases. In some circumstances this has an advantage over having a separate knowledge engineer in that the specialist knows who the users of the system are and what they need. It also shortens the path between the knowledge in the head of the specialist and its representation in the expert system. Such knowledge engineering, then, is perhaps not so much a specialist skill as one of the general life-skills of the information-based, post-industrial society.

But currently available plausible-inference software does have serious limitations. In addition to the obvious limitations in knowledge representation and processing (which we have so far avoided by restricting our effort to appropriate problems) there are other serious limitations which can be summarized by saying that much software is not sufficiently oriented towards applications. It still shows too much of its AI-research ancestry. Some characteristics of applications-oriented software not often found in current expert system software are:

a blurring of the distinction between "expert system" and "conventional" computing techniques, so that techniques are used according to their usefulness rather than their label (see below);

a good interface to other programs (of any language);

a well-engineered man–machine interface of high quality which conveys maximum useful information (this does not necessarily imply natural language input);

high run-time software efficiency;

availability on the types of computer systems that are currently being used for applications.

So it may be that we will at some stage have to change our policy of using available software and develop our own. This will probably be more a re-implementation of techniques already well-known in the AI community than a large research project.

In the meantime, the simple plausible-inference system is likely to give benefit in a number of applications. Our experience so far, constructing knowledge bases in a handful of widely differing applications (some of which cannot be described for confidentiality reasons) and discussing many others with end-users, has convinced us that plausible inference can aid the user in a number of small tasks, like:

making a choice;

analysing risks (as with the SCCES);

advising on the appropriateness of certain actions;
guiding an operator;
recognizing patterns;
simple modelling;
as well as the well-known task of diagnosing diseases or problems.

Such systems would largely be used in the consultancy role, with the benefits that were discussed in that section, but instead of simulating specialist consultants, as many of the full-blown expert systems do, with multiple knowledge representations, natural language input, etc., many of them would be small, simpler systems, rather like intelligent calculators. They would give quick, reliable answers with some form of human window in a variety of mundane tasks.

However, as any relatively new software technique becomes widely used, it will be used in more complex ways. In many published expert systems so far, this complexity has manifested itself in more flexible, more complex or multiple knowledge representations and knowledge processing methods. This is understandable, since the purpose of many such systems is not solely to solve an applications problem, but is at least partly to further AI research. The result is that most of these systems, like our small intelligent calculators, are self-contained; they stand alone.

The increasing complexity of use in I.C.I. may well take a different direction, however. In this direction the expert system will not become a larger, more complex, stand-alone system but will become half-hidden as a module in a larger system (see Fig. 1). The SCCES, for instance, might develop to become just one useful part of a computer-aided design suite for designing chemical plant and would be activated whenever SCC was to be considered. Beside it in the suite there would be other knowledge bases which the (general purpose) expert system software could employ—for other forms of SCC, for other corrosion problems, for vibration analysis, and even, we believe, for analysis of economic factors—and other kinds of programs—to route pipework, to access a materials database, etc.

When the expert system is activated it might call on other programs to perform certain calculations. For instance, it has become clear that in the SCCES, mechanisms

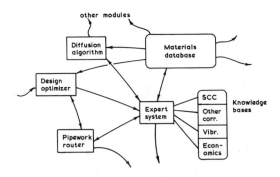

FIG. 1. The expert system half-hidden.

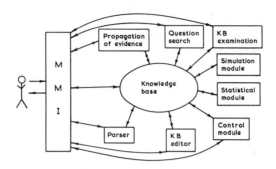

FIG. 2. The expert system totally hidden.

additional to plausible inference are needed—sometimes numerical calculations must be performed for, say, diffusion rates or the concentration of ions. Rather than require the expert system software itself to support facilities of this kind, it would be wiser simply to allow it to access other "conventional" programs or procedures written in a language more suited to such calculations. This is especially true where the procedures are either complex or already available in some software library. Or the expert system may need to access an existing database to obtain answers to some of its questions or to store results. In such suites the emphasis is shifted away from the distinctness of the computing techniques employed in expert systems to problems of interfacing with the existing, wider world which comprises the user and other software. Such considerations as communication protocols may appear trivial to those engaged in expert systems research, but they are, as with any type of software, often crucial to the success of an application.

However, not only can we see expert systems as stand-alone or as being half-hidden parts of larger suites, we can see some as being totally hidden. In contrast to the CAD suite above, where the expert system would still be a recognizable and distinct piece of software, the hidden expert system would be a distributed collection of callable procedures which perform human-like processing.

As Sandewall (1975) and others indicate, this is likely to take the form of a separate knowledge base surrounded by various modules (Fig. 2). It is the structure that thas been ; dopted by SOPHIE (Brown *et al.*, 1982), HEARSAY (Erman, Hayes-Roth, Lesser & Reddy, 1980) and many others and allows clean inter-module communication and a dump-and-restore facility, both of which are important in large applications systems.

In Fig. 2 the expert system as such has disappeared having been split into several smaller modules, each of which is responsible for carrying out one sub-task of an expert system. Alongside them are "conventional" modules such as statistical modules or optimizers, and in a well-designed system each module might make use of others. Thus the MMI module could itself be "intelligent", making use of the evidence-propagation, question-search or other expert system modules which are available in the system. The module which examines the knowledge base could form the basis of a human window on the whole system, including some of the "conventional" parts.

In this way there perhaps ceases to be "an" expert system. Instead, various expert system procedures and techniques are spread throughout the whole system, giving the benefit of their human-like processing to every part. Like the grain of wheat mentioned by Jesus Christ in the Bible (*The Gospel according to St John*, **XII**, v. 24), perhaps the expert system must "die" before it can truly bear fruit.

7. Conclusions

One relatively conventional expert system, that predicts the risk of stress–corrosion-cracking, has been described in order to illustrate a number of issues concerned with the application of expert systems. But it is of interest in being a causal model rather than a model of specific problem-solving expertise and some possible implications of this are discussed throughout the article.

It has been argued that there is a larger number of roles than originally thought, in which expert systems can be applied to real-life problems. Real benefits are offered in those roles, in some cases even when the accuracy of the knowledge base is not very high. The benefits arise from a number of advantages that expert systems have over both "conventional" computing techniques and over manual techniques.

Some limitations of expert systems are also presented. In particular, expert systems are more suited to "deep and narrow" domains of expertise, though they do have some application in "wide and shallow" domains. Here they can be useful as aids to refining knowledge in the domain.

The last section outlines part of I.C.I.'s expert systems strategy and makes the point that the much-despised simple plausible-inference system is able to give significant benefit. It also suggests that in some applications expert systems could become more hidden, as parts of larger systems, instead of standing alone. Some are likely to take the form of a central knowledge base surrounded by program modules of all kinds, both "expert system" and "conventional".

The author would like to thank the Directors of I.C.I. plc for permission to publish this paper, and his wife, Ruth, and his colleagues, J. G. Hines, B. A. Kelly and P. J. Moreland, for many constructive comments during its preparation.

References

ANDERSON, R. H., GALLEGOS, M., GILLOGLY, J. J., GREENBERG, R. & VILLANEUVA, R. (1977). *RITA Reference Manual*. Rand Corporation.
BARR, A. & FEIGENBAUM, E. A. (1982). *The Handbook of Artificial Intelligence*, vol. II. London: Pitman Press.
BROWN, J. S., BURTON, R. R. & deKLEER, J. (1982). Knowledge engineering and pedagogical techniques in SOPHIE I, II and III. In SLEEMAN, D. & BROWN, J. S., *Intelligent Tutoring Systems*. London: Academic Press.
CHECKLAND, P. (1981). *Systems Thinking, Systems Practice*. New York: Wiley.
CLANCEY, W. J. (1979). Transfer of rule-based expertise through a tutorial dialogue. *Report No. STAN-CS-769*, Computer Science Department, Stanford University.
CLANCEY, W. J. (1983). The epistemdogy of a rule-based expert system—a framework for explanation. *Artificial Intelligence*, **20** (3), 215–252 (May).
COOMBS, M. J. & HUGHES, S. (1982). Extending the range of expert systems: principles for the design of a computer consultant. *Expert Systems 82*. Technical Conference of the BCS Expert Systems Group, 14–16 September 1982, Brunel University, U.K.

DAVIS R. & LENAT, D. (1982). *Knowledge Based Systems in Artificial Intelligence*, pp. 229–490. New York: McGraw-Hill.

DUDA, R. O., GASCHNIG, J. & HART, P. E. (1979). Model design in the PROSPECTOR consultant system for mineral exploration. In MICHIE, D., Ed., *Expert Systems in the Microelectronic Age*, pp. 153–167. Edinburgh University Press.

ERMAN, L. D., HAYES-ROTH, F., LESSER, V. R. & REDDY, D. R. (1980). The HEARSAY-II speech understanding systems: integrating knowledge to resolve uncertainty. *Computing Surveys*, **12** (12), 213–253.

FEIGENBAUM, E. A. (1977). The art of artificial intelligence: 1. Themes and case studies of knowledge engineering. *International Joint Conference on Artificial Intelligence (IJCAI)*, vol. V, pp. 1014–1028.

GEORGE, F. (1980). Unpublished lecture.

ISIS (1982). *MicroExpert*. Redhill, Surrey: ISIS Systems Ltd.

JIPDEC (1981). *Preliminary report on study and research on fifth-generation computers*. Japan: Japan Information Processing Development Center.

MCALLESTER, D. A. (1980). An outlook on truth maintenance. *Memo No. 551*, AI Laboratory, Massachusetts Institute of Technology.

OLDFIELD, J. W. & SUTTON, W. H. (1978). *British Corrosion Journal* , **13**, 13.

PATERSON, A. (1981). *AL/X User Manual*. Oxford: Intelligent Terminals Ltd.

SANDEWALL, E. (1975). Ideas about management of LISP data bases. *International Joint Conference on Artificial Intelligence (IJCAI)*, vol. IV, pp. 585–592.

SLEEMAN, D. & BROWN, J. S. (1982) *Intelligent Tutoring Systems*. London: Academic Press.

SPL (1982). *SAGE*. London: Systems Programming Ltd.

STEFIK, M., AIKINS, J., BALZER, R., BENOIT, J., BIRNBAUM, L., HAYES-ROTH, F. & SACERDOTI, E. (1982). The organisation of expert systems, a tutorial. *Artificial Intelligence*, **18** (2), 135–174 (March).

STEVENS, A., ROBERTS, B., STEAD, L., FORBUS, K., STEINBERG, C. & SMITH, B. (1981). STEAMER: advanced computer aided instruction in propulsion engineering. *Report No. 4702*. Cambridge, Massachusetts: Bolt Beranek and Newman Inc.

Adapting a consultation system to critique user plans

Curtis P. Langlotz and Edward H. Shortliffe

Heuristic Programming Project, Departments of Medicine and Computer Science, Stanford University, Stanford, California 94305, U.S.A.

A predominant model for expert consultation systems is one in which a computer program simulates the decision making processes of an expert. The expert system typically collects data from the user and renders a solution. Experience with regular physician use of ONCOCIN, an expert system that assists with the treatment of cancer patients, has revealed that system users can be annoyed by this approach. In an attempt to overcome this barrier to system acceptance, ONCOCIN has been adapted to accept, analyze, and critique a physician's own therapy plan. A *critique* is an explanation of the significant differences between the plan that would have been proposed by the expert system and the plan proposed by the user. The critique helps resolve these differences and provides a less intrusive method of computer-assisted consultation because the user need not be interrupted in the majority of cases—those in which no significant differences occur. Extension of previous rule-based explanation techniques has been required to generate critiques of this type.

Introduction

As symbolic reasoning techniques that had developed in the field of artificial intelligence (AI) matured, their potential power as problem solving tools became clear and a number of expert advice systems were developed using AI methods (Duda & Shortliffe, 1983). One of the important lessons learned from the early expert consultants is that *excellent decision-making performance does not guarantee user acceptance* (Shortliffe, 1982). This lesson has been particularly evident in the field of medicine where experience applying AI techniques has helped reveal what capabilities can make computer-based consultation systems attractive to users, and what characteristics detract from their acceptability. A consultation program's ability to explain its reasoning has been shown to be particularly important in obtaining user acceptance (Teach & Shortliffe, 1981).

In this paper we describe recent additional lessons learned through our work on ONCOCIN, an expert system used by physicians in their routine care of cancer patients. Because this system is used for data management as well as for consultations, physicians do not always require or desire advice when they use the program. This is an important distinction from conventional expert systems in which the program can typically assume that the user is specifically seeking assistance with a decision task. We believe it will be preferable to allow ONCOCIN's users to indicate their own management plans, and to monitor those plans for apparent errors, rather than to generate advice routinely. Early experience suggests that this *critiquing model†* of expert system interaction will enhance a program's acceptability for some applications.

The paper begins with a brief description of the ONCOCIN design strategy and explains how the concept of a critiquing model evolved from our recent experience

---

† To our knowledge, the term *critiquing model* was first used by Miller (1983) in his work on the ATTENDING system described later in this paper.

DEVELOPMENTS IN EXPERT SYSTEMS
ISBN 0-12-187580-6

using the system as a conventional consultation program. The program's architecture is described and we then provide a detailed description and examples of our recent experiments in adapting the system for the critiquing model. We conclude with a discussion of our plans to implement the model in the operational system as we adapt ONCOCIN to run on professional workstations rather than the main-frame machine on which it currently operates.

ONCOCIN and the critiquing model

ONCOCIN (Shortliffe *et al.*, 1981) is a medical consultation program which uses knowledge of cancer treatment protocols,† encoded in production rules, to assist physicians with therapy decisions for cancer patients. It is designed to provide excellent decision-making performance while addressing a number of acceptability issues. Thus, although our main purpose is to provide high-quality management advice, we have tried to develop a system that physicians can use directly, and that they consider both helpful and suitable for regular use.

ONCOCIN is also designed to avoid additive time demands on the already busy schedule of an oncologist (cancer specialist). Rather than requiring the physician to perform a new task, it *replaces* one that many oncologists already performed manually, namely that of filling out patient data forms. Before the introduction of ONCOCIN, clinic oncologists were routinely required to fill in a time-oriented record, called a "flowsheet", which is used to maintain patient data for analysis of the effectiveness of alternate therapy regimens. Each row in the paper flowsheet corresponds to a particular test or finding. For each patient-visit, the physician enters relevant signs, symptoms, and laboratory data in the column on the flowsheet corresponding to that visit. ONCOCIN allows these entries to be made at a computer terminal instead. Consequently, ONCOCIN's computer records function as the primary source of protocol data for that patient. This requires that the physicians use ONCOCIN for *all* visits of patients whose records are on the computer, but avoids the need (and frequent inaccuracy) of clerical transcription of flowsheet data into statistical analysis routines. The availability of on-line data also allows ONCOCIN to print out automatically a variety of paper reports which physicians were accustomed to preparing by hand. These features serve to reduce the amount of time physicians must spend performing routine tasks and have heightened ONCOCIN's acceptance by its users.

We introduced the prototype system in our outpatient oncology clinic in May 1981. Physicians were involved from the outset in the system's design, and we have received valuable additional feedback from them since its introduction. The most frequent complaint raised by physicians who use ONCOCIN is that they become annoyed with changing or "overriding" ONCOCIN's treatment suggestions. Physicians override ONCOCIN's decision if they disagree, even slightly, with the program's treatment recommendation and choose to give an alternate treatment instead. Each time this occurs, the physician is asked to provide a justification for any changes. The "override" feature allows the physician to remain the final decision-maker regarding the treatment

---

† Protocols are detailed documents that specify alternate therapies to be compared in a formal experiment. In particular, they specify guidelines for delaying or modifying treatments according to patient response. They also define the data that need to be collected in order to provide an adequate basis for judging the merits of the alternate therapy plans.

given to a patient. However, if the user must frequently override ONCOCIN, this can be irritating and time-consuming, particularly when the changes are minor and reflect slight dosing adjustments for patient convenience or to increase compliance with the regimen.

It occurred to us that if the consultation system were modified to monitor and critique the therapy plans proposed by users, it could conduct a consultation in a much less disruptive manner. In addition, it would allow the doctor routinely to suggest treatment first and thereby to remain more actively involved in the consultation process. The educational role of the system would also be heightened because a critiquing system could not only point out the differences between its own recommendation and the physician's but it could also help the physician make an informed choice between the two by explaining the differences.

Little would be gained by such an approach if the consultation program were to enter into a lengthy critique of *every* plan entered by the user. However, if there were only minor differences between the physician's proposed therapy and the optimal therapy determined by the consultation system, the physician would not need to be bothered by the expert decision-making portion of the program. Only when a significant disagreement occurred would ONCOCIN interrupt to explain the problem it had noted.

An additional appeal of the critiquing approach was that it would facilitate clearer explanations when they were needed. The user's analyzed therapy plan could be used to focus the discourse because it would provide an accurate indication of the areas of knowledge of importance to the user. The goals of the explanation dialogue would thereby be more clear than if the user had simply asked a general question of the system. Measures of what is important to a user are difficult to determine in other ways (Wallis & Shortliffe, 1982).

The requirement that a system explain differences between its own recommendation and that of the user places the development of the critiquing model within the area of explanation research. Because disagreements occur often, even among experts (Yu *et al.*, 1979), it is not surprising that many system users consider an ability to explain advice the single most important feature for computer-based clinical consultation systems. This is certainly the case for diagnostic or treatment advice systems designed for physicians (Teach & Shortliffe, 1981). When disagreements between the computer and user occur, explanation can provide assurance that the computer's reasoning is logical and that unexpected advice is appropriate (Scott, Clancey, Davis & Shortliffe, 1977). Some existing expert systems (Clancey & Letsinger, 1981; Shortliffe 1976; Swartout, 1981) have tried to address this demand by providing explanations or tutorial facilities during and after a consultation. However, the therapy critiquing system developed for MYCIN (Clancey, 1977) and Miller's ATTENDING program (Miller, 1983) are, to our knowledge, the only previous programs designed to resolve conflicts between a computer-based consultant's advice and a plan preferred by the program's user. Clancey's critiquing system offered only a composite critique for each proposed therapy recommendation. It was unable to give an analysis in varied levels of detail, or to select appropriate sub-parts of MYCIN's reasoning process to explain to the user. ATTENDING critiques a preoperative anesthetic plan using knowledge of anesthesia in a decision network of anesthetic procedures and their associated risks. Risks associated with alternate procedures in the network are used to generate a prose analysis which compares the relative risks and benefits of the user's proposed plan with those of alternate strategies. As with Clancey's system, ATTENDING offers a summary analysis; users are unable to ask for analyses of particular parts of their proposed plans.

By enabling ONCOCIN to analyze and critique physicians' therapy plans, we have tried to prevent the irritations of frequent physician overrides. The system allows the doctor to routinely suggest treatment first,[†] then interrupts only if it detects a significant difference between its own recommendation and that of the doctor. It is able to relax the constraints under which it identifies conflicts with the doctor's plan; consequently, the user can prescribe without being interrupted a therapy plan that is clinically acceptable within a tolerance range of the computer's plan. The physician retains the initiative, but ONCOCIN's advisory capabilities are still available whenever they are needed.

ONCOCIN's system architecture

ONCOCIN is currently implemented on a time-shared computer which allows us to structure the system as two separate programs running in parallel: the *Interviewer*, a display program which manages the terminal used by the doctors, and the *Reasoner*, a rule-based AI program which makes decisions about the treatment of a patient according to the data it receives from the *Interviewer* (Gerring, Shortliffe & van Melle, 1982).

THE INTERVIEWER

ONCOCIN allows the doctor to enter patient data directly into the computer through the *Interviewer*, a specially designed, display-oriented interface that mimics the format of the familiar paper flowsheet. After seeing each patient, the doctor uses the computer terminal, controlled by the *Interviewer*, and communicates via a simple control keypad that moves the cursor to arbitrary locations on the computer flowsheet for data entry. Figure 1 shows an example of what the terminal screen looks like during a typical session with the physician. The *Interviewer* screen is divided into sections. At the bottom of the screen is a row of "soft key" descriptors. These correspond to keys labeled with roman numerals located across the top of the specialized keyboard. The *Interviewer* assigns different meanings to these keys depending on which portion of the flowsheet is currently displayed. In Fig. 1, for example, when the key labeled "VI" (corresponding to "CHANGE OLD DATA") is pressed, the system enters a mode which allows the physician to change or update data from previous visit dates. At another time in the session, key VI might have a different label and effect.

The central portion of the *Interviewer* screen is the "flowsheet" section. In Fig. 1 the disease activity section of the flowsheet is shown (labeled "Dz Activity"). Each column is labeled at the top with a patient visit date, and each row is labeled on the left with the name of the particular finding or text result to which it corresponds. The right-most column corresponds to the current visit date for which the physician is entering data.

---

† It should be noted that nothing in the critiquing process *requires* that the physician enter a proposed recommendation before seeing the computer's recommendation. In fact, some users might prefer to use the computer's recommendation as a guide, and to enter only their proposed changes.

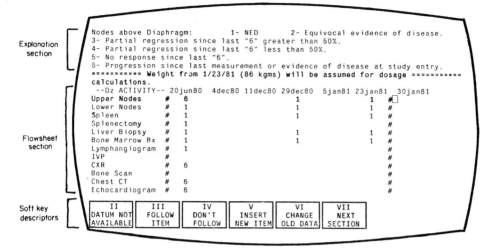

FIG. 1. A sample explanation from ONCOCIN. In the explanation section of the screen is an explanation for the "Upper Nodes" disease activity datum on the flowsheet. The square box indicates the cursor location on the screen.

Above the computer-based flowsheet is the "explanation" section. Explanations are displayed there automatically according to the location of the cursor on the flowsheet. Two kinds of explanations may be displayed: explanations of what data the system is expecting (during the data entry process) or brief explanations of the

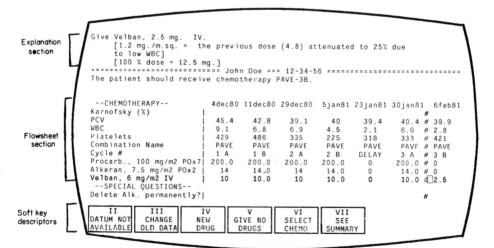

FIG. 2. A display on which the therapy recommendations of the *Reasoner* have been displayed. In contrast to Fig. 1, this time the *explanation section* shows the system's justification of the therapy advice corresponding to the user-controlled cursor location (box).

program's reasoning (when the cursor is aligned with part of a recommendation). In the data entry display of Fig. 1, for example, the cursor rests in the row corresponding to the "Upper Nodes" item. Since this is an item for which the physician is expected to provide the response, the explanation field provides more detailed information about the item, and makes clear what types of responses are expected. In this case the explanation indicates that the response entered for the "Upper Nodes" disease activity datum may range between *1* and *6*: *1* signifies "No Evidence of Disease" (NED), *2* signifies equivocal evidence of disease, etc.

As the interaction continues, the physician uses the specialized keypad to move the cursor down the flowsheet. When the physician reaches the place on the computer flowsheet where he or she would normally have written the therapy plan on the paper flowsheet, it has already been filled in with the treatment recommendation derived by the *Reasoner* using rules that represent both the protocol knowledge and the judgmental knowledge acquired from our collaborating experts. As shown in Fig. 2, the second kind of explanation is shown for those data items filled in by the *Reasoner*. In this example, since the cursor rests in the row for the *Reasoner*'s suggested dosage of the drug "Velban", a brief justification of the computer's therapy recommendation is given.

THE REASONER

The *Reasoner* uses production rules that encode specific knowledge of chemotherapy protocols as well as general strategies of oncology chemotherapy. These rules, together with data about the patient being treated, are used to make therapy recommendations using both data- and goal-driven reasoning.[†] Each datum, as it is received from the *Interviewer*, is passed to the *Reasoner* which can use it when formulating a therapy recommendation. In a typical consultation, ONCOCIN determines whether the patient should receive chemotherapy on this visit, have therapy delayed, or have the current cycle aborted.[‡] If the patient can be treated, then ONCOCIN determines the appropriate dose for each drug in the chemotherapy to which the current patient is assigned. The proper dose is determined by first deciding whether each drug should be omitted. If a drug should not be omitted, it determines the appropriate level of dose attenuation (i.e. dose reduction) or escalation, if any. The result of the reasoning process is a comprehensive chemotherapy treatment plan.

The critiquing process

We describe here our work in developing a prototype critiquing module for ONCOCIN. The capability is not yet part of the system being used in the clinic because we intend to refine it and make use of graphical input and display as ONCOCIN is transferred to run on personal LISP machines. Thus we plan to use the critiquing model as the primary mode of interaction in the new hardware environment to which our physician–users will be introduced over the next few years. Until then ONCOCIN will continue to suggest therapy for all patients as was described above.

---

† See Shortliffe *et al.* (1981) for a more detailed description of ONCOCIN's control structure.

‡ Cancer chemotherapy is often given in "cycles" in which drugs are given on two days a week apart (the "A" and "B" cycle); then a long pause of 3 or 4 weeks is taken before the next cycle begins. A cycle is "aborted" if, after an "A" cycle, no "B" cycle is given because of toxic reactions to therapy.

In the first step of the prototype critiquing process, the physician enters patient data on the computer flowsheet. Using the data, ONCOCIN formulates a recommendation. Instead of showing the recommendation to the user, however, it is withheld until after he or she has entered a proposed recommendation. The critiquing process then uses an evaluation process which systematically compares the physician's plan to the one formulated by ONCOCIN. Explanations of clinically significant differences are generated if the physician requests them. Once satisfied, the physician can (1) accept ONCOCIN's recommendation, (2) modify ONCOCIN's recommendation, or (3) choose to follow his or her original recommendation. The following sections describe in detail the critiquing techniques used.

HIERARCHICAL PLAN ANALYSIS

It has been shown that comparison with an expert can be an effective evaluation method (Burton & Brown, 1979). Thus, after the physician's recommendation is obtained, the therapy plan is evaluated by comparing it with the computer's recommendation. The comparison process uses domain-specific knowledge of the components of a recommendation and their interdependencies. For example, ONCOCIN's recommendation hierarchy shown in Fig. 3 indicates that decisions about radiation therapy

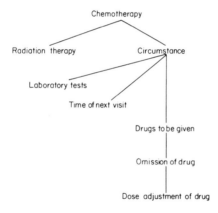

FIG. 3. ONCOCIN's hierarchy of therapy planning knowledge used in the recommendation evaluation process.

are dependent on the choice of chemotherapy, and that decisions about the dose adjustments of a drug are dependent on whether that agent is to be omitted. These types of hierarchical relationships frequently exist in clinical medicine [see Patil, Szolovitz & Schwartz (1981) for another example]. At each level of the hierarchy, ONCOCIN determines whether two analogous components of the computer's and physician's recommendations differ in a clinically significant way using an evaluation procedure specifically designed for that component of a recommendation. For example, in ONCOCIN, we obtained drug-specific dosage tolerances from an expert oncologist;

they are used to determine whether a clinically significant difference exists between two different drug doses.†

Hierarchical plan analysis is an evaluation process designed to find the most general set of differences which completely account for the significant disagreements between two therapy plans. An analysis begins at the most general component in the hierarchy. Analogous components from each recommendation are compared using the corresponding evaluation procedure. Thus, an evaluation begins with the comparison of chemotherapies. If they are in agreement, the sub-components of *Chemotherapy* (*Radiation Therapy* and *Circumstance*‡) are each evaluated. If a clinically significant difference is found in *all* such sub-components, then all differences have been accounted for and explanations will be generated for these sub-components. If a difference occurs for only *some* sub-components, they are noted as significant differences for which explanations need to be generated, but the offspring of the remaining branches are explored. The process is continued, investigating further sub-components, until each path from the top-most component either ends in a significant difference or has been explored to its fullest extent.

Consider a case of MOPP chemotherapy in which the doctor decides to delay therapy but ONCOCIN concludes that therapy should be given (a significant difference between the *Circumstance* parameters of the two recommendations). The analysis procedure begins by comparing the chemotherapies being used in each therapy recommendation and finds no significant difference (see Fig. 4). Since none is found, the offspring of *Chemotherapy* in the hierarchy are investigated. No significant difference is found between the *Radiation Therapy* components of the two recommendations, but the *Radiation Therapy* node has no offspring. Thus the analysis continues with *Circumstance*. Here the evaluation procedure finds a difference. Consequently, there is no need to perform any comparisons on the offspring of *Circumstance*. Since there are no other offspring of the *Chemotherapy* node of the hierarchy, the analysis process stops and the difference in *Circumstance* will be explained.

Now suppose for the same patient that ONCOCIN decides that the drug prednisone should be omitted, but the physician proposes to give treatment without omitting that drug (see Fig. 5). In this case the analysis begins as it did in the previous example, but since no difference in *Circumstance* is found, the analysis continues to all the offspring of *Circumstance* and all of their offspring without finding a significant difference. However, when the analysis process considers *Omission of a drug* for the drug prednisone, a significant difference is detected. Thus, no comparison will be made for the offspring of that node, namely the dose adjustment of prednisone. Instead the conclusion that prednisone be omitted will be explained. This corresponds to our intuitions about what should be explained in this case. ONCOCIN's conclusion that the drug should not be given is the relevant fact, not what led ONCOCIN to recommend any particular dose for the drug.

---

† A challenging topic for future research is to model the expert's knowledge that allows him to calculate reasonable tollerances for the various drugs used in cancer therapy. This would allow ONCOCIN to determine dynamically the significance of dosing differences and to better explain its basis for deciding that the physician's plan involves an inappropriately high or low dose. Currently it can only quote the tolerance provided by our collaborating expert.

‡ *Circumstance* is the parameter whose value determines whether the patient should receive chemotherapy on this visit, have therapy delayed, or have the current cycle aborted.

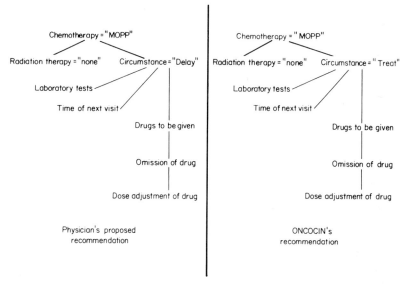

FIG. 4. A schematic representation of the hierarchy overlay at termination after it has found significant differences between *Circumstance* elements of the two recommendations. The bold lines indicate those portions of the hierarchy that were searched before a significant difference was noted (see text).

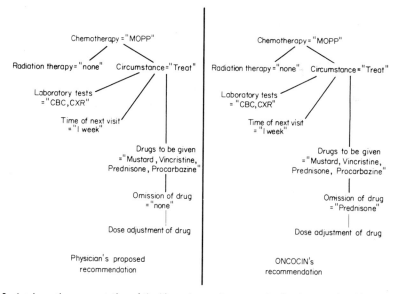

FIG. 5. A schematic representation of the hierarchy overlay at termination in a case in which a significant difference between the two recommendations was not noted until the issue of drug omission was considered. Again, the bold lines indicate those portions of the hierarchy that were searched before the discrepancy was noted.

Although we do not consider proposed recommendations with multiple differences in our examples above, it is important to note that the plan analysis process will find multiple differences if they are independent of one another. In contrast, since the analysis of any given branch in the hierarchy stops as soon as a significant difference is found, specific differences that occur as a result of a more general difference will not be considered separately. In particular, as the hierarchy is processed, each evaluation procedure can make the assumption that there is agreement between all components of the recommendations which occur above it in the hierarchy. For example, it makes little sense to evaluate the dosing of a particular drug recommended by the user when it was given under the assumption that a different chemotherapy was being used. Similarly it makes little sense to critique the dosing of a particular drug given by the doctor when the reason that a disagreement occurred is that ONCOCIN thought therapy should be delayed (different *Circumstance*) and did not recommend to give any drugs at all.

CRITIQUE GENERATION: EXPLANATION OF DIFFERENCES

For a rule chaining system like ONCOCIN, an appropriate way to explain the program's conclusions is to allow the physician to explore a record or "trace" of the computer's reasoning in detail. Such a rule trace represents a series of links between findings (data) and treatments (goals). Teach (1983) has shown that some physicians give significantly higher evaluations to explanations which make specific links between symptoms and diagnoses. It is presumed that the explanation of ONCOCIN's reasoning trace will help to inform users where their reasoning may have differed from that of the computer.† There are at least four modes of explanation used by consultation systems; our discussion will focus on the last.

1. *Explanations during a consultation* (Scott *et al.*, 1977; Swartout, 1981). The user, when asked to provide data in such systems, can respond instead with a question of the form "Why do you want to know?". The system's answers provides some indication of how the datum will be used in the reasoning process.

2. *Post-consultation explanation in response to specific questions from the user* (Scott *et al.*, 1977; Swartout, 1981). Post-consultation explanation methods often employ simple parsers to allow the user to ask about the knowledge used and actions taken in a particular consultation session. For example, the user might ask how the system made certain conclusions, how it used a piece of information, or what decision it made about some subproblem.

3. *Tutorial explanation* (Clancey, 1979). A tutorial dialogue may involve explanations of the *strategies* used to solve a problem. A model of the user and a means to evaluate the user's performance are important elements of a system which generates useful tutorial explanations.

4. *Explanation as a critique* (Miller, 1983). A critiquing system uses knowledge about the structure of the problem and its possible solutions to find important differences between the user's proposed solution to a problem and a computer-generated "expert" solution. These differences are used to structure the resulting explanation, called a critique.

---

† Underlying support and strategic knowledge will also heighten the quality of the explanations (Clancey, 1983).

REPRESENTATION AND CONTROL STRUCTURES NECESSARY FOR EXPLANATION

In order to generate post-consultation explanations, it is important that a system maintain a record of its reasoning and that the system have methods for explaining any part of that record. This eliminates the need to re-run the consultation for each explanation (Scott *et al.*, 1977).

ONCOCIN stores a *justification* each time the value of a parameter is concluded. The justification may be the rule which caused the conclusion, or some other indication of the data structure from which the information was obtained. In addition, each time a rule is used, the situation in which it was applied is recorded. ONCOCIN stores five major types of justifications with its conclusions, and has mechanisms for explaining each.

RULEXXX: (e.g. "RULE046") indicates the rule that was invoked to cause the parameter's value to be concluded (e.g. the attenuated dose of a drug). The context in which the conclusion was made is contained on the property list of the rule as described above.

DEFAULT: indicates that one or more rules were tried in order to find the value of the parameter, but none succeeded, so the default value of the parameter was assumed[†] (e.g. it is concluded that a drug should be given if no rules are found which can conclude that the drug should be omitted).

ASKED: signifies that the parameter's value was obtained by asking the user (e.g. white blood cell or platelet count).

ALWAYSCONCLUDE: signifies the parameter's value was obtained from the domain-specific protocol knowledge (e.g. the maximum dose of a drug which can be given to a patient).

HISTORY: signifies this value was obtained from the patient's medical record (e.g. the type of tumor for which the patient is being treated). This information is entered the first time the patient is seen and need not be asked again since it is stored in the patient's file.

When a physician needs an explanation for parameters whose values are justified by the HISTORY or ALWAYSCONCLUDE flags, special purpose (domain dependent) routines are used. For example, if the user asks about a parameter whose justification is ALWAYSCONCLUDE, one such routine allows ONCOCIN to respond:

```
ONCOCIN concluded the 100% dose is 6 mg/m2 according to
    protocol.
Would you like to see in more detail what the protocol
    specifies for the current chemotherapy?
```

Patient history data and the specfic instructions of the protocol do not lend themselves well to further explanation.

DOMAIN INDEPENDENT EXPLANATIONS

Many rule-based systems determine the values of parameters using three primary means: by rules, by default, and by asking the user. Consequently, we have maintained the domain independence of the explanations for these types of conclusions.

† A default rule is always tried last if no other rule has succeeded, its condition always succeeds, and its action concludes the default value. This feature can be implemented in MYCIN and other EMYCIN systems using "self-referencing" rules (Shortliffe, 1976).

Whenever a rule is used to conclude a parameter, it is stored as the justification for that parameter. A translation method used by Scott *et al.* (1977), called *static translation*, can be used to generate English text from the machine-readable form of such a rule. The statically translated form of one of ONCOCIN's production rules is:

```
To determine the dose attenuation due to low WBC for
    Nitrogen Mustard in MOPP, for Cytoxan in MOPP or for
    Cytoxan in C-MOPP:
If: white blood count (in thousands) is between 3.5 and 4
Then: Conclude that the dose attenuation due to low WBC
    is 70%
```

The static translation includes all of the situations in which the rule might apply: "for Nitrogen Mustard in MOPP, for Cytoxan in MOPP or for Cytoxan in C-MOPP". In answer to questions of the form "Why did you conclude that the dose attenuation due to low WBC is 70%?" the critiquing system uses a refinement of this technique called *dynamic translation*. The dynamically translated form of a rule contains specific values for the patient under consideration, and includes only the context in which the rule was actually used. A dynamically translated form of the same rule is:

```
To determine the dose attenuation due to low WBC for
    Nitrogen Mustard in MOPP:
Since white blood count (in thousands) (3.6) is between
    3.5 and 4
It is possible to conclude that the dose attenuation due
    to low WBC is 70%
```

In contrast, the dynamic translation includes only the situation in which the rule applied to the patient under consideration, "for Nitrogen Mustard in MOPP", and it includes the actual value of the white blood count which caused the rule to succeed, "(3·6)". Since the rule has succeeded, the condition of the rule is known to be true and the action has been taken.

Default conclusions can be explained using a similar rule-related method, except that the translation of no single rule can give a complete explanation. ONCOCIN makes a default conclusion when no rules succeed to conclude a given parameter. The fact that no rules succeed indicates that at least one of the conditions failed in each one that was tried. To explain a default conclusion, ONCOCIN therefore looks through all the rules which *could* have concluded a value for the parameter in question, and finds the part of the condition which caused each such rule to fail. These failed conjuncts can be collected, negated, and dynamically translated to form a concise explanation of why the default value was used:

```
ONCOCIN did not conclude that the patient has just had
significant radiation because the site of radiation is
not:
    1) Pelvic,
    2) Mantle,
    3) Para-aortic or Inverted-Y, or
    4) Total-nodal, Sub-total-nodal or Whole-body
```

Parameters whose values which were obtained from the user are explained simply by saying so:

```
ONCOCIN concluded platelet count (in thousands) is 200
    because that is the value that was entered. Would you
    like to change it?
```

EXPLANATION STRATEGIES

The analysis of the physician's proposed therapy plan allows for focusing the explanation dialogue around certain conclusions. An explanation is needed only for those parameters whose values are significantly different from the corresponding ones in the user's proposed therapy plan. But it would be unacceptable to print a detailed explanation of the entire chain of reasoning for a given conclusion because it is unlikely that the user wants *all* of the information contained in a complex chain with many branches. Without a detailed model of the user and a dynamically updated store of contextual information, it is difficult to find heuristics to determine what branch of reasoning should be explained first, and in what manner. ONCOCIN avoids these difficult issues by allowing users to structure explanations to their own needs using an agenda-based system.

Explanations are printed using natural language translations generated directly from the machine-readable form of the rules.† As the translation process occurs, each parameter in the rule being translated is added to a list of parameters relevant to the current line of explanation called the *agenda*. After each explanation of a parameter the user can choose a parameter in the agenda to be explained next. As each new explanation is generated, the parameters in it are added to the agenda and thus become available for further explanation. The user can select another parameter in the agenda to see further explanations. In this way, the reasoning tree can be traversed in an arbitrary order according to the needs of the user (who will generally have no knowledge of ONCOCIN's control structure, parameter names, or specific rules).

ONCOCIN uses three generic types of explanations to explain to the user how it concluded the values of parameters in the agenda. Each type is designed to answer one of the following questions.

1, Why did ONCOCIN conclude what it did?
2. Why did ONCOCIN not conclude something *other* than the value that it did.
3. Why did ONCOCIN *not* conclude some specific value for a parameter?‡

An example

The following is an excerpt from a critique in which the difference between the physician's and ONCOCIN's dose of Nitrogen Mustard was determined to be clinically significant.

```
ONCOCIN recommended:
Give Nitrogen Mustard, 8.0 mg. IV.
    [4.2 mg./m.sq.=attenuated to 70% due to low WBC]
```

† See Scott *et al.* (1977) for a complete description of the translation mechanism used.
‡ This is treated as a special case of 2.

Whereas you recommended:
Give Nitrogen Mustard, 4.0 mg. IV.
    4.2 mg./m.sq.=attenuated to 35%]

> *[To begin, a simple summary of the differences*
> *is printed.]*

Would you like me to explain why ONCOCIN concluded that it did?
    Yes†

> Agenda =(ATTENDOSE)

> *[For demonstration purposes, the agenda*
> *will be printed at each iteration. This*
> *is not normally shown to the user.]*

ONCOCIN concluded that the current attenuated dose is
    attenuated to 70% due to low WBC because for Nitrogen
    Mustard in MOPP for patients with all lymphomas:
Since the patient has not received chemotherapy and the blood
    counts do warrant dose attenuation, it is possible to
    conclude that the current attenuated dose is 100% dose
    attenuated by the minimum of the dose attenuation due to low
    WBC and the dose attenuation due to low platelets.

> Agenda =(ATTENDOSE NORMALCOUNTS WBCATTEN PLTATTEN
> CHEMODOSE)

> *[All parameters that occurred in the rule*
> *that was just printed are added to the agenda.]*

Which of ONCOCIN's conclusions would you like explained further?
0) none
1) the current attenuated dose is attenuated
      to 70% due to low WBC
2) the blood counts do warrant dose attenuation
3) the dose attenuation due to low WBC is 70%
4) the dose attenuation due to low platelets is 100%
5) the 100% dose is 6

Choose a number> 2
ONCOCIN concluded the blood counts do warrant dose attenuation
    because the dose attenuation due to low WBC is not 100%

> *[Because the conclusion that the blood counts warrant dose*
> *attenuation was made by default, its justification is obtained by*
> *translating the negation of any conjuncts which were false. These*
> *conjuncts caused rules which might have concluded otherwise (i.e.*
> *that the blood counts do not warrant attenuation) to fail. In this*
> *case there is only one such conjunct: the fact that the dose*
> *attenuation due to low WBC is not 100%.]*

> Agenda =(ATTENDOSE NORMALCOUNTS WBCATTEN PLTATTEN
>   CHEMODOSE)

> *[No new parameters were introduced in*
> *the previous explanation, so the agenda*
> *remains the same.]*

---

† Those portions of the critique dialogue typed by the user are shown underlined. Commentary is shown in italics.

Which of ONCOCIN's conclusions would you like explained further?
0) none
1) the current attenuated dose is attenuated
    to 70% due to low WBC
2) the blood counts do warrant dose attenuation
3) the dose attenuation due to low WBC is 70%
4) the dose attenuation due to low platelets is 100%
5) the 100% dose is 6

Choose a number> 3

ONCOCIN concluded that the dose attenuation due to low WBC is
    70% because for Nitrogen Mustard in MOPP:
Since white blood count (in thousands) (3.6) is between 3.5 and 4
It is possible to conclude that the dose attenuation due to low
    WBC is 70%

> *[Because the rule is translated dynamically,*
> *the actual value of the white blood cell*
> *count appears, and only the context which*
> *caused the rule to be tried is printed.]*

> Agenda=(ATTENDOSE NORMALCOUNTS WBCATTEN PLTATTEN
> CHEMODOSE WBC)

> *[The parameter "WBC" occurred in the last*
> *explanation, so it is added to the agenda.]*

Which of ONCOCIN's conclusions would you like explained further?
0) none
1) the current attenuated dose is attenuated
    to 70% due to low WBC
2) the blood counts do warrant dose attenuation
3) the dose attenuation due to low WBC is 70%
4) the dose attenuation due to low platelets is 100%
5) the 100% dose is 6
6) white blood count (in thousands) is 3.6

Choose a number> 6

ONCOCIN concluded white blood count (in thousands) 3.6 because
    that is the value that was entered.
Would you like to change it? no

> *[If users ask about a parameter that they*
> *entered, they are given an opportunity to*
> *change the value.]*

> Agenda = (ATTENDOSE NORMALCOUNTS WBCATTEN PLTATTEN
> CHEMODOSE WBC)

Which of ONCOCIN's conclusion would you like explained further?
0) none
1) the current attenuated dose is attenuated
    to 70% due to low WBC
2) the blood counts do warrant dose attenuation
3) the dose attenuation due to low WBC is 70%
4) the dose attenuation due to low platelets is 100%
5) the 100% dose is 6
6) white blood count (in thousands) is 3.6

```
Choose a number> 0
```

Would you like to change your prescribed dose of Nitrogen
    Mustard? . . .

Future work

Before the critiquing adaptation shown here becomes a regular part of ONCOCIN, it is important that both its performance and acceptability be thoroughly tested. Just as it must be demonstrated that a consultation system gives excellent advice before it is brought to a clinical setting, a critiquing system must demonstrate the ability to detect deviations from the protocol whenever experts consider them clinically significant.

Similarly, since adapting ONCOCIN to critique therapy plans involves significant changes in the computer–physician interface, it is important that these changes be tested by physicians. Physicians' feedback will be particularly important in developing a program that will conveniently accept their proposed therapy plans or their proposed changes to ONCOCIN's therapy plans.

Other ongoing work is aimed at generating explanations that will depart from simple rule translations and be more like normal discourse. Enhancing the quality of explanations is an acceptability problem whose solutions are likely to be found in AI techniques. We envision more general explanations, in which underlying support knowledge and a more detailed model of the user are used to generate statements that summarize complex chains of rules. These explanations will emphasize parts of the explanation which are most likely to be of importance to the user, while condensing or omitting areas which might be highly complex or beyond the user's level of expertise using techniques similar to those suggested by Wallis & Shortliffe (1982).

Recent innovations in AI hardware play a significant role in shaping our future research. Professional workstations will soon be available at prices similar to those currently paid by physicians who buy office computer systems. These workstations are small enough to fit conveniently into an office, and they boast high resolution graphics capabilities along with the power and speed needed to run complex AI programs. We are currently transporting the *Reasoner* to one such professional workstation and we have developed an experimental workstation version of the prototype described above. The workstation critiquing module highlights on the screen those phrases in the explanation text which correspond to parameters on the explanation agenda. These highlighted phrases can subsequently be selected for further explanation by pointing to the phrase using a movable pointer controlled by a hand-held device called a *mouse*. Thus the agenda need not be reprinted before each user choice, and more information can be shown on the screen at one time. This experiment, as well as others conducted using the workstation, indicate that graphics capabilities can be an effective means to circumvent natural language issues. It is computationally less expensive to allow users to select phrases on the screen using a mouse pointer than to interpret users' free text input. Furthermore, a dexterous user of a mouse (or similar device such as a touch screen or light pen) can participate in much more rapid interaction than is possible using a conventional computer keyboard. A user is likely to be especially intolerant of a system which requires him to wait for a response, or which requires a significant amount of typing.

Conclusion

ONCOCIN has evolved in response to comments and suggestions from physicians. User complaints about the need to override system advice have led to work which has provided us with a number of important lessons about the computing tools necessary to critique therapy plans. Of particular importance are the central roles of both plan analysis and explanation techniques. Plan analysis can indicate where significant disagreements have occurred, and thus provide information about what areas are likely to be of greatest interest to the use. This kind of information is difficult to obtain in other ways. Simple plan analysis is performed in our system using a procedure which compares components parts of two recommendations using a group of domain-specific evaluation procedures. The most general set of differences which completely describes the significant disagreements between two recommendations is found. The analysis requires evaluation criteria for each part of a therapy plan, as well as knowledge about the hierarchical relationships of entities in a typical plan.

Another important lesson learned from our work with the critiquing model is that optimal critiquing requires sophisticated explanation techniques. Generating useful explanations consists of two separate problems: the development of the *capability to explain* a chain of reasoning in a variety of ways, and the development of techniques to give the program the *knowledge about the situation* to help it determine what explanation techniques to apply and on what knowledge structures to apply them. Considerable advances remain to be made in both of these areas. For instance, the applicability of the hierarchical plan analysis technique has not been tested in other problem domains. It should be easily adaptable to domains in which the hierarchical relationships on which the analysis depends occur frequently (as they do in clinical medicine). However, to be adapted for the wide variety of problems for which expert systems are used, the critiquing model may require more general methods for representing the structure of the problems and their possible solutions. Advances are also needed to develop additional levels of specificity in automated explanations. For example, we envision more general explanations in which strategic knowledge and more detailed information about the user lead to statements that summarize complex chains of rules.

Because of its advantages both for the user and as a computational device to obtain information about the user, we believe the critiquing model has considerably more utility than the conventional consultation model in building many interactive expert systems. Our work with ONCOCIN has shown that an expert system can be reconfigured in a manner which may significantly reduce the burden on the user. Instead of acting as a mechanized consultant that methodically asks for findings and renders a treatment decision which the physician must override if a disagreement exists, the expert clinical consultant becomes a silent partner in the decision-making process and only makes its opinions known when a sub-optimal therapy is proposed by the physician. For an experienced physician these interruptions will be infrequent, and thereby less disruptive. When they occur, they offer a focused analysis of where the differences lie and why they may exist. We believe that this kind of critiquing interaction will contribute to increased acceptance of expert systems by individuals who prefer to reach decisions independently when possible.

Support for this work was provided by the National Library of Medicine under Grants LM-03395 and LM-00048, and by the Office of Naval Research under contract NR 049-479. Dr Shortliffe is a Henry J. Kaiser Family Foundation Faculty Scholar in General Internal Medicine. Computing facilities were provided by the SUMEX-AIM resource under NIH Grant RR-00785. We also thank Bruce Buchanan for helpful comments on an earlier draft of this paper.

References

BURTON, R. R. & BROWN, J. S. (1979). An investigation of computer coaching for informal learning activities. *International Journal of Man–Machine Studies*, **11**, 5–24.

CLANCEY, W. J. (1977). An antibiotic therapy selector which provides for explanations. In *Proceedings of the Fifth International Joint Conference on Artificial Intelligence*, p. 858.

CLANCEY, W. J. (1979). Tutoring rules for guiding a case method dialogue. *International Journal of Man–Machine Studies*, **11**, 25–49.

CLANCEY, W. J. (1983). The epistemology of a rule-based expert system: a framework for explanation. *Artificial Intelligence*, **20** (3), 215–251.

CLANCEY, W. J. & LETSINGER, R. (1981). NEOMYCIN: Reconfiguring a rule-based expert system for application to teaching. In *Proceedings of the Seventh International Joint Conference on Artificial Intelligence*, Vancouver, British Columbia, pp. 829–836.

DUDA, R. O. & SHORTLIFFE, E. H. (1983). Expert systems research. *Science*, **220**, 261–268.

GERRING, P. F., SHORTLIFFE, E. H. & VAN MELLE, W. (1982). The Interviewer/Reasoner model: an approach to improving system responsiveness in interactive AI systems. *AI Magazine*, **3** (4), 24–27.

MILLER, P. L. (1983). Critiquing anesthetic management: the ATTENDING computer system. *Anesthesiology*, **53**, 362–369.

PATIL, R. S., SZOLOVITZ, P. & SCHWARTZ, W. B. (1981). Causal understanding of patient illness in medical diagnosis. In *Proceedings of the Seventh International Joint Conference on Artificial Intelligence*, Vancouver, British Columbia, pp. 893–899.

SCOTT, A. C., CLANCEY, W. J., DAVIS, R. & SHORTLIFFE, E. H. (1977). Explanation capabilities of production-based consultation systems. *American Journal of Computational Linguistics, Microfiche 62.* [Also to appear in BUCHANAN, B. G. & SHORTLIFFE, E. H., *Rule-based Expert Systems: The MYCIN Experiments of the Stanford Heuristic Programming Project.* Reading, Massachusetts: Addison–Wesley (in press).]

SHORTLIFFE, E. H. (1976). *Computer-based Medical Consultations: MYCIN.* New York: Elsevier/North-Holland.

SHORTLIFFE, E. H. (1982). The computer and medical decision making: good advice is not enough. *IEEE Engineering in Medicine and Biology Magazine*, **1** (2), 16–18 (guest editorial).

SHORTLIFFE, E. H., SCOTT, A. C., BISCHOFF, M., CAMPBELL, A. B., VAN MELLE, W. & JACOBS, C. (1981). ONCOCIN: An expert system for oncology protocol management. In *Proceedings of the Seventh International Joint Conference on Artificial Intelligence*, Vancouver, British Columbia, pp. 876–881.

SWARTOUT, W. R. (1981). Explaining and justifying expert consulting programs. In *Proceedings of the Seventh International Joint Conference on Artificial Intelligence*, Vancouver, British Columbia, pp. 815–822.

TEACH, R. L. (1983). Patterns of Explanation and Reasoning in Clinical Medicine. *Doctoral dissertation*, Stanford University (in preparation).

TEACH, R. L. & SHORTLIFFE, E. H. (1981). An analysis of physician attitudes regarding computer-based clinical consultation systems. *Computers and Biomedical Research*, **14**, 542–558.

WALLIS, J. W. & SHORTLIFFE, E. H. (1982). Explanatory power for medical expert systems: studies in the representation of causal relationships for clinical consultations. *Methods of Information in Medicine*, **21**, 127–136.

YU, V. L., FAGAN, L. M., WRAITH, S. M., CLANCY, W. J., SCOTT, A. C., HANNIGAN, J. F., BLUM, R. L., BUCHANAN, B. G. & COHEN, S. N. (1979). Antimicrobial selection by a computer: a blinded evaluation by infectious disease experts. *Journal of the American Medical Association*, **242** (12), 1279–1282.

Towards an understanding of the role of experience in the evolution from novice to expert

JANET L. KOLODNER

School of Information and Computer Science, Georgia Institute of Technology, Atlanta, Georgia 30332, U.S.A.

Two major factors seem to distinguish novices from experts. First, experts generally know more about their domain. Second, experts are better than novices at applying and using that knowledge effectively. Within AI, the traditional approach to expertise has concentrated on the first difference. Thus, "expert systems" research has revolved around extracting the rules experts use and developing problem solving methodologies for dealing with those rules. Unlike these systems, human experts are able to introspect about their knowledge and learn from past experience. It is this view of expertise, based on the second distinguishing feature above, that we are exploring. Such a view requires a reasoning model based on organization of experience in a long-term memory, and incremental learning and refinement of both reasoning processes and domain knowledge. This paper will present the basis for this view, the reasoning model it implies, and a computer program which begins to implement the theory. The program, called SHRINK, models psychiatric diagnosis and treatment.

1. Introduction

In considering what distinguishes a novice from an expert, we notice two major differences:

experts are more knowledgeable about their domain and
an expert knows how to apply and use his knowledge more effectively
than does a novice.

Within the Artificial Intelligence community, studies of expertise have concentrated on problems associated with the first of these. In expert systems research, the problems of (1) extracting the knowledge experts use and representing that knowledge in rules (e.g. Davis, Buchanan & Shortliffe, 1977), (2) general problem-solving strategies for applying the knowledge (e.g. Pople, 1977; van Melle, 1980), and (3) developing consultation protocols so that diagnostic questions will be presented in coherent order have been addressed (e.g. Aikens, 1980; Weiss & Kulikowski, 1979). In addition, expert systems for diagnosis and treatment have been developed in many different domains. The premise behind this work has been that the *knowledge* used for diagnosis and treatment is of primary importance (Feigenbaum, 1977), and that representing that knowledge in the form of rules makes use and explanation of a system easy. Thus, the stress has been on extracting rules from experts and building systems which from the beginning contain all or most of the compiled knowledge an expert has. This view of expertise has led to many working systems [e.g. MYCIN (Shortliffe, 1976), CADUCEUS (Pople, 1977), PUFF (Kunz, 1978), HEADMED (Heiser, Brooks & Ballard, 1978), CASNET (Weiss, Kulikowski, Amarel & Safir, 1978)] and many issues associated with representation and problem-solving have been addressed. There is

95

something vital missing, however, when this approach is taken—the ability to use and learn from experience.

Our study of expertise is concerned with the role experience plays in an expert's reasoning and how experience changes the way an expert reasons. We are addressing expertise from the second point of view given above. From this point of view, we need to find out what reasoning capabilities experts have that novices lack. Our approach to that problem is to look at the evolution of a memory from novice to expert. Two things happen in that evolution. First, knowledge is built up incrementally on the basis of experience. Facts, once unrelated, get integrated through occurrence in the same episodes. Second, reasoning processes are refined, and usefulness and rigidity of rules is learned. As a result of noticing failures and successes, and differences and similarities between "cases", the ability to deal with exceptional or novel cases is derived. Reasoning processes, the knowledge they use, and incremental changes in the organizational structure of the knowledge and the reasoning processes are all equally important. Because experience is vital to the evolution from novice to expert, experience is organized in long-term memory, and guides reasoning processes. Failures in reasoning processes guide incremental change to the memory organization and thus to the reasoning processes themselves. Successful reasoning reinforces knowledge and processes already in use.

In fact, this explanation of the development of expertise seems reasonable when considering people. When a person has only gone to school and acquired book knowledge, he is considered a novice. After he has experience using the knowledge he has learned, and when he knows how it applies both to common and exceptional cases, he is called an expert. In fact, it is not uncommon to hear experts from such diverse domains as medicine, law, and computer programming to remark that "in school they taught us everything we needed to know about . . . , but it wasn't until I got actual experience that I was really able to appreciate that knowledge and figure out how to apply it". Medical and other professional schools which recognize that experience is important have two modes of teaching—they teach facts in the classroom, and also teach how to use the facts through supervised experience (internship) in the real world. Experience serves to turn unrelated facts into expert knowledge.

The evolution from novice to expert requires introspection and examination of the knowledge used in solving problems. Thus, we see human experts interpreting new cases in terms of something with which they are already familiar (either a previous case or generalized knowledge). This implies that as an expert is having new experiences, he is evaluating and understanding them in terms of previous ones. In the process, he must also be integrating the new experience into his memory so that it too will be accessible to use in understanding a later case. Our goal is to understand how this happens. Drawing on recent research into the organization of experience in long-term memory (Kolodner, 1981*a*, *b*, 1983*a*, *b*; Schank, 1980), our long-term goal is to address the following problems.

1. What makes an expert expert?
2. What kind of processing capabilities are implied when we say somebody is an expert?
3. What kind of understanding capability does an expert have that a novice does not?
4. What are the processes that comprise expert reasoning?
5. What is the relationship between experience and expertise?

We start by considering how experience can act to turn facts into expert knowledge. One way to address this problem is to look at what psychologists have labeled "episodic" and "semantic" memory. Semantic or generic memory has been described as the facts we know, arranged in a hierarchical network. In a semantic memory, for example, "penicillin" may be defined as an "antibiotic" which is in turn defined as a "drug". "Drug" may have a property called "used to cure" which points to "diseases" or "medical problems". "Antibiotic" may have the same property with value "infection", which is itself defined as a "medical problem". Similarly, particular antibiotics may point to the particular infections they combat. Nowhere in such a memory is the fact that "last time I (i.e. the system) treated a patient with penicillin, he broke out in hives". In other words, while the facts in semantic memory may have been compiled from experience at some time, semantic memory itself is totally separate from experience. It has little operational or functional information associated with it.

Episodic memory, on the other hand, encodes experience. Concepts in an episodic memory are related to each other according to their co-occurrence in the same episode. Thus, the relationships between objects are much richer than in a semantic store. Furthermore, in an episodic memory, an object is defined by its function, i.e. its role in different operations or situations. An episodic rendition of some fact will have particular episodes associated with it, and eventually generalizations concerning how to use it and how useful it is. Thus, in an episodic memory, "antibiotic" would be defined with respect to the kind of episodes in which it has been used, namely, when serious infection arises and the patient goes to the doctor or hospital, a doctor determines the type of infection and then prescribes an appropriate antibiotic to combat it, the expected result being that the infection goes away. In addition, particular circumstances under which antibiotics were used, the results of their use, and generalizations about the appropriateness of use of a particular antibiotic for a particular set of symptoms are also recorded. Such recording, along with a method for integrating new cases into memory and traversing memory structures, allows previous similar cases or experiences to be extracted from memory and used for reasoning when a new case is encountered. Thus, in addition to generalized knowledge, cases which both support and break the rules are available to use in reasoning about a new case.

Consider again, the two distinguishing factors of an expert. The first takes only the semantic store into account and says "the expert has more knowledge than a non-expert". Traditional expert systems research (knowledge engineering) investigates that knowledge. While on the surface it seems to give us an explanation of expertise, it doesn't get at the crucial differentiator between a novice and an expert—experience. The second answer addresses that crucial difference. It implies that even if a novice and an expert had the same *semantic* knowledge (i.e. knew the same facts), the expert's experience would have allowed him to build up better *episodic* definitions of how to use it.

Addressing the problem of expertise from the first perspective encourages systems with rules and meta-rules (similar to MYCIN). Such systems are, in principle, unable to learn new facts as an expert does. In such a system, "learning" a new fact means putting a new rule into the system. As more and more new facts are "learned", the list of rules grows. The system gets slower and more control is needed. Such an approach ignores the problem of how knowledge gets *integrated* with other knowledge and *reorganized* over time. Furthermore, the difficulty of extracting rules from experts supports the contention that experts don't use rules for reasoning, but rather that they reason in some other way. Looking at the problem from the second point of view suggests that expert reasoning is done based on "cases" rather than rules.

Experts, after all, don't slow down or require more reasoning as they learn more. On the contrary, their reasoning becomes more automatic and faster. This is because they are learning how to use the knowledge they once learned as facts. Experience allows the *facts* they once learned to be related to each other through the episodes they take part in together. The factual knowledge becomes integrated with experience, and as a result more integrated with other facts. The use of similar facts in similar situations allows generalizations to be made and used where, before experience, only the facts themselves could be used. As a result of experience, for example, a doctor can learn which are the vital signs to look for first in making a diagnosis, or can surmize the kind of reaction a particular type of patient might have to a new medication. He might also learn how and when to stretch the "rules" he learned as a student. In order for this to happen, memory's organization must change as a result of experience, and each experience must contribute to memory's ongoing *reorganization*.

If we want to design a computer system which can introspect and learn incrementally over time, we must allow it to remember what it has done and provide it with feedback concerning the outcomes of decisions it has made. We must also discover the reasoning processes used in expert reasoning, how memory must be organized to access that knowledge easily, and how memory's organization changes over time. We can then build a memory with initial basic knowledge, and give the program experiences (e.g. cases to diagnose and treat) and feedback on its performance (e.g. the results of applying the treatment). Based on its initial memory organization, its reasoning rules, and its rules for reorganizing memory to incorporate new experiences, we can make it reason as an expert and learn as it deals with new cases.

While this might have seemed like a strange idea a few years ago, in light of recent work within Artificial Intelligence on episodic memory (e.g. Kolodner, 1983*b*; Schank, 1982), the idea not only seems promising, but we also have some idea of how to do it. There are at least two types of experience-based learning that we must investigate:

similarity-based generalization and
failure-driven learning.

Similarity-based generalization has previously been implemented in CYRUS (Kolodner, 1983*b*) and IPP (Lebowitz, 1983). It allows specializations of already-known concepts to be created based on similarities between items described by those concepts. It is constrained by considering only those features of new items which differ from other similar items. In this method of learning, similar differences are used to specialize and refine already-defined conceptual categories.

Failure-driven learning has been proposed by Schank (1982) and implemented by Riesbeck (1981) in a program called ALFRED. In failure-driven learning, process failures or exceptions are noted, and an attempt is made to explain them (or assign blame). If blame is assigned, an explanation of the failure is left as a marker. When a similar situation is encountered, the marker serves as an index to the failed episode. If a solution was found to the first failure situation, it can be applied to the second so that the same failure doesn't happen again. When blame cannot be assigned, a marker denoting the difference between the failed episode and normal ones is left, again serving as an index when a similar situation is encountered. In this case, the second time a similar situation is encountered, both can be examined to determine a explanation for the failure, and that solution will be applied. The explanation is recorded so that it can be applied the next time a similar situation is encountered.

In order to investigate experience-based learning, we are drawing on the previous research mentioned above and are also observing physicians as they make decisions. We are building a computer program, called SHRINK, which implements and tests our theory of expertise in the domain of psychiatry. When complete, the program will analyze new psychiatric cases based on previous cases it has seen. It will integrate new cases into its memory as it is processing them, and will build up its expertise based on generalizations it has made concerning the similarities between cases it has seen. Its expertise will also be heightened through analysis of failures in diagnosis and treatment.

In the field of psychiatry, a recently-published manual for doctors called the *Diagnostic and Statistical Manual of Mental Disorders* (DSM-III) (American Psychiatric Association, 1980) provides procedures for psychiatric diagnosis. Knowledge about symptoms and disorders are integrated into the process. Related disorders are organized in decision trees. Predominant symptoms (called primary indices) suggest entry points into those trees. The book also specifies sets of necessary, supporting, additional, and exclusion criteria for each currently-recognized psychiatric disorder. We are starting with the processes and diagnostic categories specified in that book as a model of a novice diagnostician, and giving our program experiences which allow it to reorganize that knowledge "episodically" and learn.

There are at least two important types of knowledge which can be learned by integrating experience with "book" knowledge:

1. exceptions to prescribed procedures and expected disorder manifestations are encountered and dealt with, refining the diagnostic process itself and
2. the relationships between diagnostic categories is learned, allowing better hypotheses to be made in novel situations.

To see that this is so, consider the following example (adapted from DSM-III Case Book, Case #125), which illustrates some of the processes mentioned above:

Dr X, just out of medical school, knows that severe mood disturbance signals that a patient suffers from major depression. He sees a patient who seems to show classic signs of major depression. She is 38 years old and her chief complaint is depression. In the last month, she has been depressed, suffering from insomnia, and crying. She has been aware of poor concentration and diminished interest in activities. She reports that she has been depressed since childhood because her father deserted the family when she was approximately ten. She has previously been diagnosed as having depression, and was treated in a mental hospital with antidepressants. Furthermore, she related that she was sickly as a child, has had a drinking problem, and has had a number of physical illnesses which doctors have not been able to find causes for.

Seeing that she has been treated previously for depression, that her chief complaint is depression, and that she has insomnia, poor concentration, and diminished interest, Dr X concludes that this patient is suffering from Major Depression, Recurrent, without Melancholia. He treats her with antidepressants. The antidepressants seem to work, but the woman keeps coming back complaining of additional major physical disorders. Dr X begins to think (because she should not still be complaining) that there may be some other problem which he had not accounted for.

He asks her for a further history and finds out specifically about the medical problems she has had. There have been about fifteen of them, and doctors have been unable to find organic reasons for most of them. At this point, he begins to realize that the large number of medical problems is important to consider. Going through the diagnostic process again using that symptom as the predominant clinical feature, he realizes that he should have diagnosed her for Somatization Disorder† in addition to the obvious diagnosis of Depression.

With this experience (if he has learned from it), Dr X has progressed one step in the evolution from novice to expert. As a result of this case, he should learn the following:

1. it is important to take medical history into account in choosing predominant clinical features, especially if it is excessive,
2. it is important to take unexplained medical problems into account in choosing predominant clinical features, especially if excessive,
3. depression can camouflage Somatization Disorder (she's never been diagnosed for it before), and
4. a patient who is highly depressed but who complains about medical problems may be suffering from Somatization Disorder also, and it should be checked out in addition to Depression.

Using the first and second facts, the doctor should be able to refine his rules for choosing predominant clinical features, This case should help him conclude that unexplained illnesses and medical history in general may be more important in choosing primary indices than he had thought. In addition, the next time he finds that a treatment has failed, he may be biased towards finding out facts about the patient's medical history that he hadn't known (i.e. if not knowing enough about the medical history of the patient made his diagnosis fail once, it is a factor to consider in assigning blame for failure of a later diagnosis). Using the third fact, he should be skeptical of diagnoses of Depression coming from other doctors, and will want to find out more about the medical history of a new patient before taking a previous diagnosis seriously. The fourth fact gives the relationship between Depression and Somatization Disorder, which could be helpful in diagnosing and treating later cases.

Furthermore, this case should enable the doctor to hypothesize that there is often more than one system that is primary, and that the patient's chief complaint may not be the most important symptom to consider. Current problems must be separated from long-standing problems and both must be taken into account. Later cases treated by this doctor should enable him to recognize and separate current from long-standing problems, to confirm these two hypotheses, and to learn specifically which long-standing problems and which chief complaints are likely to be significant.

---

† In Somatization Disorder, a patient manifests physical disorders for which no physical symptoms can be found. Rather, the symptoms are linked to psychological factors over which the patient has no voluntary control.

2. Episodic memory

To explain our approach to expert reasoning, we must further explain episodic memory. Episodic memory is memory for experience. It records and organizes individual episodes or events in a person's life. It also creates and records generalized episodes (i.e. types of events). To see what this means, consider the fact that some episodes are reminiscent of others. When one is "reminded" (Schank, 1980) of a previous episode, the similarities between the two episodes can be extracted to form a generalized episode. From two experiences at restaurants, for example, a generalized restaurant episode is formed which holds knowledge about the usual events that happen in a restaurant. If a doctor sees a particular set of symptoms a number of times, and if he finds that a particular drug seems to work in each of those cases, a similar type of generalized episode can be formed.

This generalization capability is important for a number of reasons. First, it is economical in terms of storage. Once a generalized episode is formed, full details of the individual episodes which support it do not have to be stored. Instead, only the features of each which differentiate it from the generalized episode need be recorded. The second use for generalized episodes is as an organizational structure. The generalized episode serves as an organizing point for storing similar episodes with respect to each other. If memory is well-organized in this way, memory search (the analog of remembering in people) can be directed to only the most relevant items. The third important use for generalized episodes is in understanding and reasoning. When a new experience is reminiscent of a generalized episode, the generalizations formed previously can be used to analyze and act on the new experience. Thus, when a doctor sees the same set of symptoms yet again, he can recognize the disease as one he has treated before and will not have to spend a lot of time reasoning about this particular case.

We call the structures that organize episodes in memory Episodic Memory Organization Packets (E-MOPs) (Kolodner, 1983*b*). An E-MOP is a generalized episode and thus organizes generalized information about the individual episodes which comprise it. It also organizes those episodes by indexing them according to their deviations from those norms. In previous research, E-MOPs have been used to organize the day to day events of important dignitaries in a long term memory. The E-MOP in Fig. 1 shows the organization of some of the events in former U.S. Secretary of State Cyrus Vance's life.

In this picture, EV4 is a meeting with Dayan about the Camp David Accords, EV2 represents a meeting with Gromyko about SALT, and EV3 is a meeting with Begin about the status of Jerusalem. Other meetings are indexed more deeply in the structure in MOP4 "diplomatic meetings about the Camp David Accords with Begin". Its structure is not shown but it, too, holds generalized knowledge about its members and indexes them according to their differentiating features. CYRUS (Kolodner, 1983*b*), the program which implements that research, also has E-MOPs representing each of the other activities in which dignitaries take part. There are four things to notice about this organizational structure.

1. It holds both generalized information and pointers to descriptions of actual events—the generalized information describes typical events organized in this MOP.

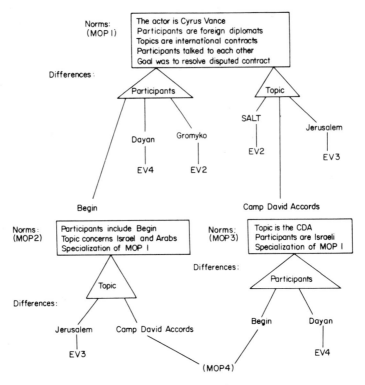

FIG. 1. "Diplomatic meetings".

2. Events and sub-MOPs are indexed by their differences from the MOP's norms.
3. It has a two tier indexing structure—the first being a feature type and the second the actual feature value.
4. The types of indices a MOP has are domain-related. In this MOP they correspond to types of features salient to diplomatic meetings.

The retrieval process which allows "reminding" of previous episodes to happen is a traversal procedure. When a new item is added to memory or a retrieval request is encountered, appropriate E-MOPs are chosen, features are extracted from the event and appropriate E-MOP indices are traversed. As a new experience is being understood and added to memory (indexed by its deviations from the norms), it collides with previous similar experiences already indexed in memory's structures. Those collisions provide remindings and trigger generalizations. This provides the basis for experience-based reasoning and learning.

2.1. MEMORY ORGANIZATION FOR PSYCHIATRIC DIAGNOSIS

In refining task expertise, two different types of knowledge must be learned or refined—domain knowledge used by the reasoning process, and the reasoning process itself. Refinement of any type of knowledge requires an explicit representation of the knowledge that can be examined and changed over time. Refinement based on experience requires a way of keeping tack of experiences associated with the knowledge that is used. Thus, explicit models of both the reasoning processes that are employed and the domain knowledge that is used must be maintained in episodic memory. We call the memory structures which represent the reasoning process PROCESS MOPs. In psychiatric disorders and treatment, the reasoning process corresponds to diagnosis, and the domain knowledge consists of diagnostic knowledge. We call the structures which organize that knowledge DIAGNOSTIC MOPs. The initial structure of each will be explained in this section. Later sections will explain how the structures change with experience.

The primary PROCESS MOP in psychiatric diagnosis is called "Psychiatric Diagnosis". The first step in psychiatric diagnosis involves examining the patient and choosing predominant clinical features (predominant symptoms). Probable disorders are chosen based on that set of data (differential diagnosis). Each is then evaluated in more detail and unlikely diagnoses are deleted (diagnostic evaluation). Failures in the diagnostic evaluation can suggest additional disorders which must be evaluated. Following is the initial (i.e. before refinement) PROCESS MOP "Psychiatric Diagnosis" (adapted from DSM-III). Note that each step of this process is also represented by a PROCESS MOP.

An individual diagnostic experience happens each time a doctor or our system diagnoses a patient. If the diagnostic process and data are similar for a group of patients, then generalizations are made about that group of patients and compiled

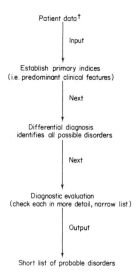

FIG. 2. Psychiatric diagnosis.

† Includes history, results of a physical examination, results of a mental status examination, and laboratory data.

paths through the process will be established. Such paths allow uncomplicated cases of particular disorders to be easily identified. Thus, there will be a compiled path with particular signs, symptoms, primary indices, differential diagnoses, etc. filled in for each commonly seen disorder. For example, a path through the process above for diagnosing a first episode of Major Depression consists of establishing that the predominant clinical feature is a mood disturbance, doing a differential diagnosis outputing "Major Depression" as the most probable disorder, and in the diagnostic evaluation finding that none of the exclusion criteria for Major Depression are present and that there have been no previous depressive or manic episodes. The output of the diagnostic evaluation will be "Major Depression, Single Episode".

The diagnosis process is driven by domain knowledge (i.e. diagnostic knowledge). This includes knowledge about particular disorders (or diagnostic categories) and symptom knowledge. The structures which record knowledge about particular disorders are called *DIAGNOSTIC MOPs*. Knowledge about a particular disorder includes its identifying signs and symptoms and how to treat it. In addition, it must include knowledge about how other disorders relate to it. "Major Affective Disorder", for example, is a general category which includes "Major Depression" and "Bipolar Disorder" (manic depression) as sub-categories. Disorders can also be related to each other through exclusion criteria (DSM-III). For example, a patient with a major affective disorder with psychotic features can be distinguished from a patient with a psychotic disorder and atypical depression by the temporal relation between their psychotic and depressive symptoms. Exclusion criteria associated with each disorder record these distinctions. In addition, each disorder category must organize individual cases of the disorder. Knowledge about individual cases is necessary in developing new treatment and diagnostic rules and in refining existing ones.

Following is some of the initial information SHRINK has about "Depressive Episodes", one of its DIAGNOSTIC MOPs. Presence of a "Depressive Episode" is necessary to diagnose almost all "Major Affective Disorders" (another of its DIAGNOSTIC MOPs).

ESSENTIAL FEATURES:
  at least 1 of:
    (1) dysphoric mood
    (2) pervasive loss of interest or pleasure in all or almost all usual pastimes and activities
SUPPORTING FEATURES:
  at least 4 of:
    (1) a significant change in appetite
    (2) a disturbance in sleep habits
    (3) psychomotor agitation or retardation
    (4) decrease in sexual drive
    (5) loss of energy or fatigue
    (6) feelings of worthlessness, self-reproach, or excessive or inappropriate guilt
    (7) complaints or evidence of diminished ability to think or concentrate
    (8) recurrent thoughts of death, suicidal ideation, wishes to be dead, or suicide attempt
  time constraint: symptoms must have been present simultaneously and for at least 2 weeks
EXCLUSION CRITERIA:
    (1) preoccupation with a mood-incongruent delusion or hallucination (indicates PSYCHO-
        TIC DISORDER).
    (2) bizarre behavior (possibly indicates PSYCHOTIC DISORDER)

ADDITIONAL CRITERIA:
  negate diagnoses of:
    (1) SCHIZOPHRENIA
    (2) SCHIZOPHRENIFORM DISORDER
    (3) ORGANIC MENTAL DISORDER
    (4) UNCOMPLICATED BEREAVEMENT
TREATMENT:
  choose a combination of:
    (1) antidepressant treatment
    (2) hospitalization
    (3) ECS therapy
    (4) psychotherapy

FIG. 3. Depressive episode.

The second type of domain knowledge memory must have is symptom knowledge—knowledge of how particular symptoms and their combinations tend to predict particular diagnostic categories. Symptoms suggest entry points into differential diagnosis decision trees, thus suggesting general diagnostic categories into which a patient may fit. Attempted suicide, for example, suggests severe mood disturbance, which suggests a possibility of Depression. Paranoid delusions are suggestive of Paranoid Disorder, Schizophrenia, or Major Affective Disorder with Psychosis. A long complicated medical history often implies Somatization Disorder.

In reasoning about a particular case, each of these types of knowledge is used in conjunction with the others. Each step of the diagnostic process is guided by either disorder or symptom knowledge. After predominant clinical features (major symptoms) are chosen, then knowledge associated with them is used to choose starting points for differential diagnosis. This happens because symptoms designate or point to the diagnostic categories with which they are most often associated. Only those categories implied by the primary symptoms are considered in initial differential diagnosis. Differential diagnosis is guided by knowledge about the relationships between diagnostic categories. Disorder knowledge (in particular, knowledge about the normal symptoms and exclusion criteria associated with particular disorders) also guides diagnostic evaluation once possible disorders have been established through differential diagnosis.

3. Incremental learning

The memory organization described so far takes care of cases which fit diagnostic criteria. In addition to dealing with normal cases, however, an expert (and any useful expert computer system) must be able to deal with unexpected or deviant cases. A patient, for example, may almost, but not quite, fit the description for a diagnostic category. Alternatively, what may initially seem like primary signs or symptoms might be masking the real problems.

In people, the evolution from novice to expert requires learning how to deal with such novel cases. One way this evolution happens is through incremental learning. Previously learned knowledge is put to use. When it fails, the reason for the failure is established. In establishing why the failure occurred, previous knowledge is "debugged". The next time a similar problem case is seen, new refined knowledge is applied to it and (hopefully) the same failure will not happen again. Similarly, when previously-learned knowledge is applied successfully, the credibility of the knowledge used must be reinforced.

In order for failures in procedure to refine previously-held diagnostic rules, memory's organization must be updated with each new experience. For that to happen in the memory structures we have defined, two things must happen when an experience deviates from the expected. First, the deviant path through appropriate PROCESS MOPs must be recorded. Second, deviant features of this case with respect to previous similar cases must be recorded in DIAGNOSTIC MOPs.

There are two reasons for recording these differences. First, they should be recorded so that when a similar deviation occurs later, the original case can be remembered. The two cases can then be examined for similarities, and generalizations based on their similarities can be used to evoke a new diagnostic category. Later, when similar cases occur, the knowledge necessary to deal with them will already be in memory. Second, a deviation signals that additional reasoning must be attempted. Extra effort can then be applied to "explain" the deviation. If an explanation is found, diagnostic procedures are updated, and that case is maintained as support of the new procedure. If no explanation is found, the deviation marks a problem that must later be resolved. Later, when a similar problem case is encountered, the marker signals that both the old case and the new one should be examined to see if between them they provide enough evidence to explain the violation. Markers in PROCESS MOPs allow incremental process changes. Markers in DIAGNOSTIC MOPs allow refinements in diagnostic categories.

Thus, there are two triggers for incremental learning which we consider:

> similarity and
> failure.

Similarity between a current and previous case triggers generalizations to be made based on similarities between the cases. Failure, on the other hand, triggers additional analysis, the results of which must then be recorded. Similarity as a trigger and the generalization process it triggers have been explained in detail in Kolodner (1983*a*, *b*) and Lebowitz (1983) and thus will not be explained here. Using the example in section 1, the remainder of this paper will concentrate on the failure-driven process.

4. Failure-driven incremental learning

To see how we aim to get the computer to learn incrementally by the failure-driven process, we will consider the example in section 1 in more detail, pointing out some of the processes and the problems involved. We have identified six processes included in failure-driven incremental learning:

1. initial decision,
2. noticing the failure,
3. assigning blame,
4. correcting the failure,
5. explaining the failure, and
6. memory update.

4.1. INITIAL DECISION

The first step in incremental learning is to follow the current procedure to come to a conclusion. In this case, that involves following the normal DIAGNOSIS procedure to make a diagnosis. In order to make the Major Depression diagnosis, the doctor must know the following.

1. "Depression", when reported by a patient, generally means severe unhappiness.
2. Severe unhappiness is a form of mood disturbance.
3. Mood disturbance suggests that the patient may suffer from Major Depression.

This knowledge allows mood disturbance to be chosen as the predominant clinical feature, which in turn allows a differential diagnosis of "Major Depression" to be made. We also assume that the doctor knows and can recognize symptoms of a Depressive Episode (as noted in Fig. 3). He recognizes three supporting symptoms for a depressive episode in addition to mood disturbance—insomnia, poor concentration, and diminished interest.

In fact, diagnosis of a Depressive Episode requires that four supporting symptoms be present in the patient. In this particular case, the doctor must use his judgement to relax that rule. Because the patient has previously been diagnosed as Depressive, he is willing to conclude that she has had a Depressive Episode even though only three symptoms are present (DSM-III Case Book, Case #125). Note that this judgement and the knowledge it uses are typical of what we want our program to learn.

One way the doctor might have learned this, and one way we can make our system learn to make judgements such as this is by having it look at a number of cases that don't quite fit the criteria but which have had previous diagnoses of depression. In analyzing the first case, the closest diagnosis it will be able to make will be Major Depression. Suppose that it treats for Major Depression and finds the treatment satisfactory. The second case in which the closest diagnosis is Major Depression, but the patient has one fewer than the prescribed number of symptoms, should "remind" the system of the previous case. If treatment is successful the second time, then a hypothesis can be made about such cases based on the features they have in common. In particular, if both patients have been previously diagnosed for depression, then a hypothesis can be made that the number of symptoms necessary to diagnose depression can be relaxed if a previous diagnosis of depression has been made on the patient. Each subsequent case with only three supporting symptoms and a previous diagnosis of depression should remind the system of this hypothesis, and each time treatment succeeds, its credibility should be strengthened. This is an example of similarity-triggered learning. Furthermore, failures should allow refinement of the hypothesis, letting the system notice which particular sets of symptoms must be present in order to relax the general rule, and which ones don't give enough support to the diagnosis. Note that such judgements may be suspect, and that if a failure occurs, this step will have to be examined as a possible reason for the failure.

The doctor concludes that the patient has Major Depression, Recurrent (she has been diagnosed for it previously), and without Melancholia (we must also assume that the doctor can recognize the symptoms of Melancholia, and that they are not present). We assume that this doctor knows that a common treatment for Major Depression is antidepressants, and therefore prescribes that treatment. As a result of the medication, the woman's unhappiness goes away, she is more interested in doing things, she sleeps well, and her concentration is improved. This is the expected successful result of the medication.

4.2. NOTICING THE FAILURE

In this case, however, there is also an additional response to the medication. The patient complains of a number of physical symptoms. This is unexpected, and must therefore be recognized as a failure, and further analyzed. Before a violation can be explained, it must be recognized. A violation of an expectation may be a reaction completely different from the expected one, or there may be varying degrees of resemblance. Alternatively, as in this case, a violation is recognized even though the expected reaction is present, since there is additional behavior that cannot be accounted for.

4.3. BLAME ASSIGNMENT

Once a failure is noticed, its cause must be found. This problem, "assigning blame" for expectation violations or failures, is hard both for people and for computers. In this case, there are two things to explain. First, causes for the new problem must be found. This is the process we call "blame assignment". Second, the doctor must find out whether the failure could have been prevented (i.e. what, if anything, failed in the diagnosis or treatment process?). The second process we call "explaining the failure".

There are three possible causes for the patient's new complaint:

(a) the medicine is producing side-effects,
(b) the patient has developed a new physical ailment independent of the medication, or
(c) the patient is imagining her disorders.

To find out what the problem really is, each of these possibilities must be investigated. The first is checked by considering whether the patient's ailments have been previously encountered as side-effects of the particular drug she is taking. Both the doctor's previous experience prescribing that drug and the medical literature must be checked to find that out. We assume that her ailments do not correlate with the drug's known side-effects. Note, however, that although this particular side-effect might not have been reported before, if no other explanation for the patient's complaints is found, then they could be considered a side-effect of the medicine.

Considering the next possibility, appropriate diagnostic tests should be performed on the patient. After taking further history, performing a physical examination, and doing screening tests, if no organic reason for the illness is found, then the doctor should be "reminded" of the patient's previous medical history. This should happen because this experience is similar to the previous medical experiences the patient has reported to the doctor—she has a history of physical illnesses for which doctors have been unable to find causes. This should lead the doctor to hypothesize that the patient is imagining the disorders, to take a further history, and to check the possibility of Somatization Disorder.

4.4. REMINDING

Note that there are many types of "reminding" that need to be done in the processing we are suggesting. The reasoner (whether human or computer) should be reminded of previous cases similar to the one it is currently working on so that knowledge gleaned from those cases can be used to reason about the new case. Such reminding is illustrated above in our explanation of how the knowledge necessary to relax a diagnosis rule could be acquired. In addition, the reasoner must be reminded of episodes associated with the case it is currently working on, so that it can notice patterns of behavior associated with the current patient.

Reminding must also allow a third type of knowledge to be remembered—symptom knowledge. Recall that when the doctor realizes that the patient is probably imagining her symptoms and that she has a long history of unexplained physical disorders, he hypothesizes that she might be suffering from Somatization Disorder. In order for him to make that hypothesis, he must be reminded of the following piece of symptom knowledge (which, in fact, he already knew).

4. Excessive unexplained physical ailments in females implies a Somatization Disorder.

A question to ask at this point is why the doctor can remember this rule now and didn't remember it when making the initial diagnosis. Our claim is that knowledge only becomes available when there is a way to direct memory processes toward it. He did not remember this rule initially because he did not choose unexplained physical ailments as a predominant clinical feature. Instead, he was focussing on the then current problem. Because he was focussing entirely on the mood disturbance, only knowledge associated with mood disturbances was accessible for reasoning. Only when attention is directed to the patient's physical disorders and medical history as possible clues to her illness, does the knowledge associated with those symptoms become available. The memory organization must support such reasoning.

4.5. CORRECTING THE FAILURE

Once the reason for an unexpected event is found, it can be corrected. Using the symptom knowledge listed above in rule 4, the doctor can now hypothesize that the patient has a Somatization Disorder in addition to Depression. He does a differential diagnosis and diagnostic evaluation based on that hypothesis and finds that her past medical history does support the hypothesis.

4.6. EXPLAINING THE FAILURE

At this point, the doctor has corrected his initial mistake. In order to learn from it, he must figure out whether and why he made the mistake. In the general domain of diagnosis and treatment, there are four possibilities:

(a) the diagnosis was wrong, and therefore the treatment is unsatisfactory,
(b) the diagnosis was right, but the treatment was not appropriate to the diagnosis,
(c) the diagnosis was right, but the treatment did not work, and
(d) the treatment and diagnosis were right, but something new has come up.

In finding the cause of the new complaint, the doctor has already found that an additional diagnosis had to be made, so the original diagnosis could have been wrong (a). On the other hand, he might not have had the necessary information initially to make this diagnosis. Therefore, (d) could also be the case.

To distinguish between the two of these, the doctor must decide whether he had the necessary information in the beginning to make the correct diagnosis. The key to making that decision is figuring out where in the diagnosis there might have been a problem and how it could have been corrected. Diagnosis consists of a number of processes and, in general, a reasoning failure may happen during any of them.

In this case, however, there is a direct route to finding the initial failure. Having corrected the mistake, the diagnostician has the crucial piece of information that allows him to figure out where in the process he went wrong. He knows that the sympton knowledge in rule 4 would have been necessary to diagnose Somatization Disorder initially. Furthermore, he knows that symptom knowledge of this sort is part of the initial diagnosis process—establishing primary indices.

His error, then, was in choosing predominant clinical features. Once he knows where in the process the error occurred, he must determine whether he had enough information initially to include Somatization Disorder in his original differential. He did have this knowledge, since the patient had already reported having been sickly and having had a number of illnesses for which no organic causes could be found. He concludes that he should have paid attention to that initially, and marks "unexplained physical disorders" and more generally "medical history" as patient features to which he should pay more attention in the future.

4.7. MEMORY UPDATE

As a result of this experience, memory must be changed, integrating the new episode with previous knowledge. In addition to concluding that medical history is an important clinical feature to consider in diagnosis, something more specific is learned—the relationship between Somatization Disorder and Depression. This will be represented in a number of ways in episodic memory. First, there will be markers or indices associated with Major Depression. One way this case differed from normal Major Depressive cases is that the patient had a large number of previous medical illnesses. That patient feature will be one of the features which index this individual case among the Major Depression cases diagnosed and treated. If another case comes up in which the patient is depressive and also has had a number of previous illnesses, then this

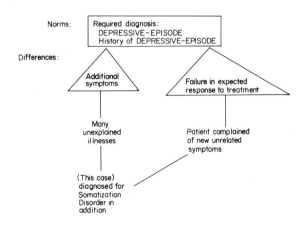

FIG. 4. Recurrent Major Depression.

case, which included an additional diagnosis of Somatization Disorder, will be remembered. This case will also be indexed as one in which the patient began complaining of other illnesses after treatment for Depression seemed successful. If another similar case comes up, the doctor or system can be reminded of this case. That should cause him to wonder whether there is a previous medical history which he had not elicited from the patient, and if the second patient also has Somatization Disorder. Episodic memory will contain similar markers associated with Somatization Disorder relating it to Depression.

Memory update adds indices to the DIAGNOSTIC MOP "Recurrent Major Depression" as illustrated in Fig. 4. Note that before this case, the norms were in place, but no experiential knowledge was recorded.

In addition, the diagnostic process itself must be marked taking this failure and its explanation into account. The particular PROCESS MOP which must be marked in this case is "Find Primary Indices" (FPI), the first step of "Diagnose" (see Fig. 2). The naive FPI PROCESS MOP is as in Fig. 5.

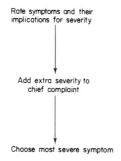

FIG. 5. Find primary indices.

After analyzing this case the markers given in Fig. 6 must be added.

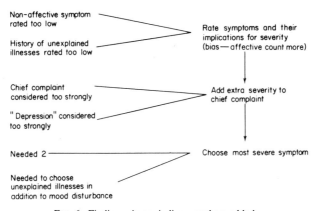

FIG. 6. Finding primary indices, markers added.

5. SHRINK, the program

In working on this project, we are both examining protocols and implementing a program, called SHRINK, which diagnoses and learns by the processes explained above. Currently, SHRINK knows about Major Depression and diagnoses normal cases of Single-Episode Depression and Manic Depressive Syndrome. Its diagnosis process is the one illustrated in Fig. 1. Figure 2 shows the knowledge that allows SHRINK to recognize Depressive Episodes. The following is an example of SHRINK making a diagnosis. The case is adapted from Case #163 of the *DSM-III Case Book*.

```
    >(diagnosis P-1)

SHRINK Diagnostic Program - Version 82-1
     Artificial Intelligence Project
     Georgia Institute of Technology

--- Patient Background and Information ---

The patient is Ms. 163, a female, 25 years old

Recently, Ms. 163 has been exhibiting the following behavior:
     attempted suicide by CARBON-MONOXIDE which caused roommates to
        take patient to emergency-room
     has had a #SIGNIFICANT# #DECREASE# in state S-WEIGHT
     has had a #SIGNIFICANT# #DECREASE# in state S-MENTAL-FOCUS
     has had a #SIGNIFICANT# #DECREASE# in drive D-SLEEP
     is less than #NORMAL# on the S-EMOTIONAL scale

Ms. 163's personal mental background and family mental background are:
     *not-given*

--------- End of Patient Record ---------

**** Begin Diagnostic Impression ****

Attempting to decide Ms. 163's predominant clinical symptom.
The predominant clinical feature is:
     attempted suicide by CARBON-MONOXIDE
Attempted suicide implies the following starting points in the
     DD-MOOD-DISTURBANCE

Performing differential diagnosis using the starting indices just found...
...Differential diagnosis completed.
     Possible diagnoses are: (MAJOR-DEPRESSION)

**** End Diagnostic Impression ****

**** Begin Diagnostic Evaluation ****

Considering MAJOR-DEPRESSION.
     Matching for essential features for MAJOR-DEPRESSION.
         Attempting deagnosis of SINGLE-EPISODE-MAJOR-DEPRESSION
             Matching for essential features of SINGLE-EPISODE-MAJOR-DEPRESSION.
                 Attempting diagnosis of DEPRESSIVE-EPISODE
                     Matching for essential features for DEPRESSIVE-EPISODE.
                     Essential features of DEPRESSIVE-EPISODE matched
                     Matching for supporting-symptoms for DEPRESSIVE-EPISODE.
                     Matching for exclusion-criteria of DEPRESSIVE-EPISODE.
```

The patient's symptoms indicate that there are
no exclusion criteria for DEPRESSIVE-EPISODE present.
However, their absense might have occured from omission of data.
For each of these exclusion criteria, please verify its absense:
...
Thank you. SHRINK continues...

Matching for additional-criteria of DEPRESSIVE-EPISODE.
Attempting to negate diagnosis of SCHIZOPHRENIC-DISORDER
Attempting to negate diagnosis of PARANOID-DISORDER
Attempting to negate diagnosis of ORGANIC-MENTAL-DISORDER
No alternate diagnoses can be found
Adding DEPRESSIVE-EPISODE to the diagnosis chain.

Essential features of SINGLE-EPISODE-MAJOR-DEPRESSION matched
Matching for supporting-symptoms for SINGLE-EPISODE-MAJOR-DEPRESSION
Matching for exclusion-criteria of SINGLE-EPISODE-MAJOR-DEPRESSION
Matching for additional-criteria of SINGLE-EPISODE-MAJOR-DEPRESSION
Adding SINGLE-EPISODE-MAJOR-DEPRESSION to the diagnosis chain.

Essential features of MAJOR-DEPRESSION matched
Matching for supporting symptoms for MAJOR-DEPRESSION
Matching for exclusion-criteria of MAJOR-DEPRESSION
Matching for additional-criteria of MAJOR-DEPRESSION

Adding MAJOR-DEPRESSION to the diagnosis chain.

Symptoms confirm MAJOR-DEPRESSION

**** End Diagnostic Evaluation ****

Ms. 163 suffers from:
(MAJOR-DEPRESSION SINGLE-EPISODE-MAJOR-DEPRESSION DEPRESSIVE-EPISODE)

End of SHRINK Run

In addition to making diagnoses, the machinery for learning by experience is being put into place. The first example we are working on is the one in section 1.

Our emphasis right now is on coming up with reasonable processes, examples, and memory structures, and not on accuracy. As the processes and memory structures become more well-defined, accuracy will become important. As SHRINK evolves, it is possible that it could become a diagnostic consultant. There are a number of additions that will have to be made to the system before that can happen, however. First, accuracy is important in such a system. More extensive disease knowledge, a natural language capability, and the ability to explain how it came to its conclusions are a few more examples of what must be added. In addition, before being used as a decision support system, it would have to be human factored to fit smoothly into standard clinical practice.

6. Implications and discussion

This paper has presented a new approach to the problem of understanding and automating expert reasoning. We take human expert reasoning as a model and attempt to both understand the human model and implement a computer program which copies it. Our long-range goal is to understand expertise. Our shorter-range goal is to understand expertise in the domain of medical reasoning, particularly psychiatry.

The theory of expertise is based upon previous research into organization of events in a long term conceptual memory (Kolodner, 1983*b*). That research described retrieval and organizational strategies for an episodic memory. Organizational structures corresponded to different types of events. In a medical domain, those structures correspond to situations in the daily life of a medical practitioner. PROCESS MOPs, representing explicitly the reasoning process used in diagnosing and treating any case, and DIAGNOSTIC MOPs, representing domain-specific diagnostic knowledge, are the organizational structures we use to store and represent a doctor's experiences. Organizing experiences in these structures allows incremental learning and refinement of the diagnostic process itself and the knowledge it uses.

There are a number of advantages and disadvantages to the approach we are taking in terms of building expert consultation systems. The first major advantage of our approach is that it is geared toward exceptional cases or novel situations, exactly those kinds of situations which a clinician in need of further advice cannot deal with himself.

In particular, such a system handles exceptional cases very nicely. The general case is stored as a diagnostic category. In any but the exceptional cases, the general knowledge associated with the diagnostic category is used for diagnosis and treatment. As exceptions are encountered (e.g. the Somatization case above where normal depressive treatment was inadequate), they are indexed off of the diagnostic category according to their differentiating features. If an explanation of the exception has been made, it is stored along with the exceptional case. When a new case is reminiscent of a previous exceptional case, knowledge about the previous case can be used to deal with the new case. When an exception has been encountered and dealt with successfully a number of times, it evolves into a new diagnostic category with its own specialized diagnostic and treatment rules. The general rules associated with the original diagnostic category do not change, however, unless an exception becomes the generalized case.

Another advantage of this approach is that it deals with both experiential knowledge and facts in the same way. Both are stored in the same structures and organized identically. This means that both are equally accessible and both can be used in reasoning. An implication of this organization is that it is amenable to new information from any source. Both new treatments and methods of diagnosis discovered through experience and those learned through journal articles or from others can be added to memory in the same way. The same processes used for reorganizing memory due to failure in experience can be used to reorganize memory based on new information acquired from elsewhere. Of course, as in people, only through experience will all the implications of such knowledge be learned and added to memory.

There is another consideration related to memory reorganization which is important to note. New information, whether acquired through experience or by some other means, should not be believed immediately. It should be possible, instead, to assign credibility to a new piece of knowledge and to reinforce or degrade that credibility over time. New knowledge based on experimentation and presented in a journal by a highly credible source, for example, might be assigned higher initial credibility than a new hypothesis that the system has made on the basis of one or two experiences. Similarly, if the source of a new diagnostic procedure or treatment is of low credibility, it can be noted in memory, but with low credibility attached. Doing this, both experientially-discovered and journal-discovered knowledge can be processed by the same algorithms. Assigning credibility is important to the process that attempts to explain failures. A piece of knowledge with low credibility is more suspect in trying to assign blame for a failure than one with higher credibility.

If we are aiming towards building expert consultation systems, this approach has the disadvantage of being highly complex. Algorithms for diagnosis and treatment are not based on any hard and fast rules. Rule-based expert systems (e.g. MYCIN) have the advantage of having a fairly simple algorithm. This makes them easy to understand by non-computer scientists. Because they are based on rules, it is also fairly simple to have them report the reasoning they went through to deal with a case. Our approach, based largely on learning, is not as simple. It may be difficult to verify a new generalization the system has made, perhaps making the system's whole operation suspect. Another challenge associated with making a system such as this acceptable as a diagnostic consultant is giving it the capability of explaining itself. This is important since acceptance by clinicians is dependent on being able to follow a program's rationale.

In the long run, we think that these will not be limitations. In fact, in developing a system based on human reasoning, we hope to be developing a system which can explain itself easily to experts and which experts will find easy to understand. It should not have the problem, as traditional rule-based systems do, of asking the human diagnostician questions he was not expecting, since it should be going through a process of diagnosis similar to what the diagnostician would be doing himself if he were an expert in the field. Thus the explanation problems should be equivalent to those found when an expert in one field attempts to explain his conclusions to an expert in another related field.

This report has pointed out how experience aids in developing the expertise necessary for expert reasoning. It has also introduced a computer program based on these ideas. The research and the program are still in a state of infancy. Nevertheless, we see this approach as having a great deal of promise both in terms of implementing expert computer systems and in helping us to understand the cognitive processes underlying expertise. The system we are proposing, which learns from its own experiences and from the experiences of others, should ultimately be able to diagnose and treat illnesses, and also to keep up with new practices and treatment as a good human expert does.

Dr Robert M. Kolodner of the Atlanta VA Medical Center has been the informant for the project. He has provided his expert knowledge of psychiatric diagnosis, going through numerous examples pointing out when he and others were going by the rules and when their expertise and experience allowed the rules to be stretched.

Keith McGreggor has been the programmer for this effort. In order to write the program, he has had to read between the lines of *DSM-III* to figure out the algorithm implied. Through his programming effort, he has been able to point out holes in the processes prescribed by *DSM-III* (e.g. it doesn't say how to choose primary indices). He has also developed representations for the disorders and symptoms the system knows about.

This material is based on work supported in part by NIH-BRSG Grant No. 5 S07 RR 07024-16 and by NSF Grant No. IST-8116892.

References

AIKENS, J. S. (1980). Prototypes and production rules: A knowledge representation for computer consultations. *PhD. thesis*. Technical Report, Department of Computer Science, Stanford University, Palo Alto, California.

AMERICAN PSYCHIATRIC ASSOCIATION (1980). *Diagnostic and Statistical Manual of Mental Disorders* (3rd Edition). Washington, D.C.: American Psychiatric Association.

DAVIS, R., BUCHANAN, B. G. & SHORTLIFFE, E. H. (1977). Production rules as a representation for a knowledge-based consultation program. *Artificial Intelligence.* **8**, 15–45.

FEIGENBAUM, E. A. (1977). The art of artificial intelligence: themes and studies of knowledge engineering. In *Proceedings of the International Joint Conference on Artificial Intelligence, 1977*, pp. 1014–1029.

HEISER, J. F., BROOKS, R. E. & BALLARD, J. P. (1978). Progress report: a computerized psychopharmacology advisor. *Proceedings of the Eleventh Collegium Internationale Neuro-Psychopharmacologicum*, Vienna, Austria.

KOLODNER, J. L. (1981*a*). Organization and retrieval in a conceptual memory for events. In *Proceedings of the International Joint Conference on Artificial Intelligence*, Vancouver, B.C., Canada, pp. 227–233.

KOLODNER, J. L. (1981*b*). Knowledge-based self-organizing memory. In *Proceedings of the IEEE International Conference on Cybernetics and Society*, Atlanta, Georgia, pp. 289–295.

KOLODNER, J. L. (1983*a*), Maintaining memory organization in a dynamic long term memory. *Cognitive Science*, **7** (4) (to appear).

KOLODNER, J. L. (1983*b*). *Retrieval and Organizational Strategies in Conceptual Memory: A Computer Model*. Hillsdale, New Jersey: Lawrence Erlbaum, Associates (forthcoming).

KUNZ, J., FALLAT, R., McCLUNG, D., OSBORN, J., VOTTERI, B., NII, H., AIKENS, J., FAGAN, L. & FEIGENBAUM, E. (1978). A physiological rule-based system for interpreting pulmonary function test results. *Report # HPP-78-19*, Computer Science Department, Stanford University, Stanford, California.

LEBOWITZ, M. (1983). Generalization from natural language text. *Cognitive Science*, **7** (1), 1–40.

POPLE, H. E. (1977). The formation of composite hypothesis in diagnostic problem solving: an exercise in synthetic reasoning. In *Proceedings of the Fifth International Joint Artificial Intelligence Conference*, Boston, Massachusetts, pp. 1030–1037.

RIESBECK, C. K. (1981). Failure-driven reminding for incremental learning. In *Proceedings of International Joint Conference on Artificial Intelligence, 1981*, pp. 115–120.

SCHANK, R. C. (1980). Language and memory. *Cognitive Science*, **4** (3), 243–284.

SCHANK, R. C. (1982). *DYNAMIC MEMORY: A Theory of Learning in People and Computers*. Cambridge: Cambridge University Press.

SHORTLIFFE, E. H. (1976). *Computer-Based Medical Consultations: MYCIN*. New York: American Elsevier.

SPITZER, R. L., SKODOL, A. E., GIBBON, M. & WILLIAMS, J. B. W. (1980). *DSM-III Case Book*. Washington, D.C.: American Psychiatric Association.

VAN MELLE, W. (1980). A domain independent system that aids in constructing consultation programs. *Report #STAN-CS-80-820*, Computer Science Department, Stanford University, Stanford, California.

WEISS, S. M. & KULIKOWSKI, C. A. (1979). EXPERT: A system for developing consultation nodes. In *Proceedings of the Sixth International Joint Conference on Artificial Intelligence*, pp. 942–947.

WEISS, S. M., KULIKOWSKI, C. A., AMAREL, S. & SAFIR, A. (1978). A model-based method for computer-aided medical decision-making, *Artificial Intelligence*, **11**, 145–172.

Strategic explanations for a diagnostic consultation system

Diane Warner Hasling, William J. Clancey and Glenn Rennels

Heuristic Programming Project, Computer Science Department, Stanford University, Stanford, California 94305, U.S.A.

This article examines the problem of automatic explanation of reasoning, especially as it relates to expert systems. By *explanation* we mean the ability of a program to discuss what it is doing in some understandable way. We first present a general framework in which to view explanation and review some of the research done in this area. We then focus on the explanation system for NEOMYCIN, a medical consultation program. A consultation program interactively helps a user to solve a problem. Our goal is to have NEOMYCIN explain its problem-solving strategies. An explanation of strategy describes the plan the program is using to reach a solution. Such an explanation is usually concrete, referring to aspects of the current problem situation. Abstract explanations articulate a general principle, which can be applied in different situations; such explanations are useful in teaching and in explaining by analogy. We describe the aspects of NEOMYCIN that make abstract strategic explanations possible—the representation of strategic knowledge explicitly and separately from domain knowledge—and demonstrate how this representation can be used to generate explanations.

1. Introduction

The ability to explain reasoning is usually considered an important component of any expert system. An explanation facility is useful on several levels: it can help knowledge engineers to debug and test the system during development, assure the sophisticated user that the system's knowledge and reasoning process is appropriate, and instruct the naive user or student about the knowledge in the system (Scott, Clancey, Davis & Shortliffe, 1977; Davis, 1976; Swartout, 1981*a*).

The problems in producing explanations can be viewed in a framework of three major considerations: epistemologic issues, user modelling, and rhetoric. This section discusses what we mean by each of these and reviews work done in each area.

1.1. EPISTEMOLOGIC ISSUES

The foundation of any explanation is a model of the knowledge and reasoning process to be explained. The explanation work that we characterize as epistemological is concerned with *the knowledge that is required to solve a problem* and *the aspects of problem-solving behavior that need to be explained*. In attempting to emulate human problem-solving activities [such as electronic trouble-shooting (Brown, Burton & deKleer, 1982)], researchers found that existing models of human reasoning were too limited to support robust problem-solving and explanation. Thus one key aspect of research in this area is the study and formalization of the reasoning process in terms of the structure of knowledge and how it is manipulated. For example, in examining causal rationalizations and explanations, deKleer & Brown (1982) discovered the problems of modelling causal processes precisely so they are powerful enough to solve problems people can solve, as well as intuitive enough for people to understand. Similar

117

DEVELOPMENTS IN EXPERT SYSTEMS
ISBN 0-12-187580-6

studies are underway for physics problem-solving (Chi, Feltovich & Glaser, 1981) and medical diagnosis (Patil, Szolovits & Schwartz, 1981; Pople, 1982).

Another aspect of this work is the design of a representation language for formalizing a model of reasoning in a computer system. Shortliffe (1976) and Davis (1976) use a simple framework of goals and inference rules to direct a medical consultation; the translation of these rules constitutes the explanation of the inference procedure. Clancey (1981) explores the issue of representing each type of knowledge separately and explicitly in order to convey it clearly to a student. Swartout (1981*b*) uses domain principles and constraints to produce a "refinement structure" that encodes the reasoning process used in constructing the consultation program. In all cases, the task in designing these systems is to represent knowledge and reasoning in a well-structured formalism that can be used to solve problems (perhaps in compiled form as in Swartout's system) and then examined to justify the program's actions.

1.2. USER MODEL

Given an idea of the knowledge needed to solve the problem and a representational framework, a model of the user can be used as a step in determining what needs to be explained to a particular person. The basic idea is to *generate an explanation that takes into account user knowledge and preferences*, often based on previous user interactions and general *a priori* models of expertise levels. The modelling component produces this picture of the user.

For example, Genesereth (1982) takes the approach of constructing a user plan in the course of an interaction to determine a user's assumptions about a complex consultation program. In ONCOCIN, Langlotz & Shortliffe (1983) are able to highlight significant differences between the user's and system's solutions by first asking the user to solve the problem, a common approach in Intelligent Tutoring Systems. In GUIDON, Clancey (1979) uses an "overlay model", in which the student's knowledge is modelled as a subset of what the expert knows. In BUGGY, Brown & Burton (1980) compiled an exhaustive representation of errors in arithmetic to identify a student's addition and subtraction "bugs".

1.3. RHETORIC

Once the content of an explanation has been determined, there is the question of how to convey this information to the user. Rhetoric is concerned with *stating the explanation so that it will be understandable.* It is here that psychological considerations (for example, the need for occasional review to respect human limitations for assimilating new information) are also examined. In STEAMER (Williams, Hollan & Stevens, 1981), Stevens explores the medium of explanation by using a simulation of a physical device, a steam propulsion plant, to produce graphic explanations supplemented with text. Choosing the appropriate level of detail (that is, pruning the internally generated explanation) has been considered by Swartout (1981*a*) and Wallis & Shortliffe (1982).

Explanations, like all communication, have structural components. For example, BLAH (Weiner, 1980) structures explanations so that they do not appear too complex, taking such things as embedded explanations and focus of attention into account. For TEXT, McKeown (1982) examined rhetorical techniques to create schemas that encode aspects of discourse structure. The system is thus able to describe the same information in different ways for different discourse purposes. In GUIDON, Clancey (1979) developed a set of discourse procedures for case method tutorial interactions. The most trivial form of structure is syntax, a problem all natural language generators must consider. At the opposite extreme some programs can produce multiparagraph text (Mann *et al.*, 1981).

2. Motivation for strategic explanations in NEOMYCIN

2.1. NEOMYCIN AND STRATEGIES

The purpose of NEOMYCIN is to develop a knowledge base that facilitates recognizing and explaining diagnostic strategies (Clancey, 1981). In terms of our framework for explanation, this is an epistemological investigation. The approach has been to model human reasoning, representing control knowledge (the diagnostic procedure) explicitly. By *explicit* we mean that the control knowledge is stated abstractly in rules, rather than embedded in application specific code, and that the control rules are separate from the domain rules.† In contrast to Davis' (1980) use of metarules for refining the invocation of base-level rules, NEOMYCIN's metarules choose among lines of reasoning, as well as among individual productions. Thus the metarules constitute a strategy in NEOMYCIN's problem area of medical diagnosis.

A *strategy* is "a careful plan or method, especially for achieving an end". To *explain* is "to make clear or plain; to give the reason for or cause of".‡ Thus in a *strategic explanation* we are trying to make clear the plans and methods used in reaching a goal, in NEOMYCIN's case, the diagnosis of a medical problem. One could imagine explaining an action in at least two ways. In the first, the specifics of the situation are cited, with the strategy remaining relatively implicit. For example, "I'm asking whether the patient is receiving any medications in order to determine if she's receiving penicillin". In the second approach, the underlying strategy is made explicit; "I'm asking whether the patient is receiving any medications because I'm interested in determining whether she's receiving penicillin. *I ask a general question before a specific one when possible*". This latter example is the kind of strategic explanation we want to generate. The general approach to solving the problem is mentioned, as well as the action taken in a particular situation. Explanations of this type allow the listener to see the larger problem-solving approach and thus to examine, and perhaps learn, the strategy being employed.

Our work is based on the hypothesis that an "understander" must have an idea of the problem-solving process, as well as domain knowledge, in order to understand the solution or solve the problem himself (Brown, Collins & Harris, 1978). Specifically, research in medical education (Elstein, Shulman & Sprafka, 1978; Benbassat & Schiffman, 1976) suggests that we state heuristics for students, teaching them explicitly how to acquire data and form diagnostic hypotheses. Other AI programs have illustrated the importance of strategies in explanations. SHRDLU (Winograd, 1972) is an early program that incorporates history keeping to provide WHY/HOW explanations of procedures used by a "robot" in a simulated BLOCKSWORLD environment. The

---

† See Clancey (1983*a*) for discussion of how diagnostic procedures can be captured by rules and still not be explicit.

‡ *Webster's New Collegiate Dictionary*.

procedures of this robot are specific to the environment; consequently, abstract explanations such as "I moved the red block *to achieve preconditions of a higher goal*" are not possible. CENTAUR (Aikins, 1980), another medical consultation system, explains its actions in terms of domain-specific operations and diagnostic prototypes. Swartout's (1983*b*) XPLAIN program refers to domain principles—general rules and constraints about the domain—in its explanations. In each of these programs, abstract principles have been instantiated and represented in problem-specific terms.

NEOMYCIN generates strategic explanations from an *abstract* representation of strategy. In contrast with other approaches, this strategic knowledge is completely separate from the domain knowledge. This general strategy is instantiated dynamically as the consultation runs. Thus when the program discusses the problem solution, it is able to state a general approach, as well as how it applies in concrete terms.

2.2. MYCIN'S EXPLANATION CAPABILITY

MYCIN is a rule-based medical consultation system designed, in part, so that it could explain its reasoning to a user. A consultation is run by backchaining through the applicable rules, asking questions when necessary. The reasoning employed by the system is explained in terms of this goal chain. There are two basic explanation commands: WHY is used to ascend the chain, discussing subsequently higher sub-goals, while HOW descends the chain explaining how sub-goals were or could be achieved. Figure 1 is an excerpt of a consultation in which the program is explaining its line of reasoning.

Here the current question, "Does the patient live in a crowded environment?" by rule 533 would imply "Is Neisseria-meningitidis causing the meningitis?". This explanation is a good example of the large amount of information embedded in the rules. The clause ordering in the rule implicitly specifies a strategy of top-down refinement—the system is to determine that the patient's infection is meningitis before trying to explore the possibility of bacterial meningitis. This highlights the disease hierarchy also obviously present in the expert's model: meningitis (clause 1) is more general than bacterial meningitis (clause 3), which is the parent of Neisseria-meningitides (the conclusion). The answer to the second WHY illustrates one of MYCIN's "task rules", used to direct the consultation at the highest level. Note the implicit procedure, perhaps apparent to the experienced physician, of gathering initial information, obtaining medical history and physical examination information, and then considering laboratory data.

Although this is a very flexible mechanism which accurately portrays what the system is doing, it has several limitations. For example, in Fig. 1 MYCIN cannot explain why it establishes that the infection is meningitis [3.1] before it determines if the meningitis is bacterial [3.3]. As indicated above, a strategy of top-down refinement of diseases is being followed. Much of the information that went into writing the rules, including justification, ordering, and planning, is either lost or made implicit and thus cannot be explained. This inexplicable information is, in essence, a large part of the strategy employed to do the diagnosis.

2.3. DESIGN CRITERIA

In determining what NEOMYCIN should explain and how it should be explained, we used several design criteria.

42) Does Mary live in a crowded environment?
** WHY

[i.e. WHY is it important to determine whether PATIENT-1 does live in a crowded environment?]

[3.0] This will aid in determining the organisms (other than those seen on cultures or smears) which might be causing the infection.
It has already been established that
  [3.1] the infection which requires therapy is meningitis, and
  [3.2] organisms were not seen on the stain of the pending csf culture (CULTURE-1), and
  [3.3] the type of the infection is bacterial
Therefore, if
  [3.4] Mary does live in a crowded environment
then
    there is weakly suggestive evidence (.2) that Neisseria-meningitidis is one of the organisms (other than those seen on cultures or smears) which might be causing the infection
Author: Yu
  [RULE533]

[back to question 42 . . .]
** WHY

[i.e. WHY is it important to determine the organisms (other than those seen on cultures or smears) which might be causing the infection?]

[4.0] . . . in order to determine the therapeutic regimen of PATIENT-1.
  If
    [4.1] information has been gathered about organisms isolated from this patient, organisms noted on smears of this patient, negative cultures of this patient, suspected infections without microbiological evidence, current drugs of this patient and prior drugs of this patient, and
    [4.2] an attempt has been made to deduce the organisms which require therapy, and
    [4.3] you have given consideration to organisms (other than those seen in cultures or smears) which might be present
  then
    determine the best therapy recommendation from among the drugs likely to be effective against the significant organisms, or indicate that no therapy is required at this time
  [RULE092]

[back to question 42 . . .]
** NO

FIG. 1. Sample MYCIN explanations.

Explanations should *not presuppose any particular user population.* The long-range goal of this research is to use NEOMYCIN as the foundation of a teaching system. At that point the strategic explanations developed here will be used to teach the strategy to students to whom it might be unfamiliar. Techniques used to generate explanations should be flexible enough to accommodate a model of the user.

Explanations should be *informative*; rule numbers or task names are not sufficient.

Explanations should be *concrete or abstract*, depending upon the situation. Thus it must be possible to produce explanations in either form. This should facilitate understanding both of the strategy and how it is actually applied.

Explanations should be *useful for the designer, as well as the end user* of NEOMYCIN. The vocabularies of computer science and an application domain, such as medicine, are different in many ways. People tend to be most comfortable with the vocabulary of their field; the system should have the flexibility to accommodate a user-dependent choice of terminology.

Explanations should be *possible* at the *lowest level of interest*; the "grain level" should be fine enough to permit this. To allow for use in debugging, we chose the level of rules and tasks as our lowest level. Higher level explanations can later be generated by omitting details below the appropriate level.

The following explanation of strategy is an example of how we try to satisfy these criteria in NEOMYCIN. Note how the explanation is abstract, more similar to a MYCIN "task rule" (e.g. rule 92 in Fig. 1) than a domain rule (e.g. rule 533).

> 17) Has Mary been hospitalized recently?
> ** WHY
>
> [i.e. WHY is it important to determine whether Mary
> has been hospitalized recently?]
>
> [21.0] We are trying to round out the diagnostic
> information by looking generally into past
> medical history and by reviewing systems.
>
> There are unasked general questions that can help us
> with the diagnosis.

3. How strategic explanations are possible—The NEOMYCIN system

MYCIN (Shortliffe, 1976), the precursor of NEOMYCIN, is unable to explain its strategy because much of the strategic information is implicit in the ordering of rule clauses (Clancey, 1983*a*). In NEOMYCIN, the problem-solving strategy is both explicit and general. This section provides an overview of the representation of this strategy in NEOMYCIN, since this is the basis for our strategic explanations. Other aspects of the system, such as the disease taxonomy and other structuring of the domain knowledge, are described in Clancey & Letsinger (1981).

NEOMYCIN'S strategy is structured in terms of *tasks*, which correspond to metalevel goals and subgoals, and metalevel rules (*metarules*), which are the methods for achieving these goals. The metarules invoke other tasks, ultimately invoking the base-level interpreter to pursue domain goals or apply domain rules. Figure 2 illustrates a portion of the task structure, with metarules linking the tasks. The entire structure currently includes 30 tasks and 74 metarules. This task structure represents a general diagnostic problem-solving method. Although our base-level for development has been medicine, none of the tasks or metarules mention the medical domain. As a result the strategy might be ported to other domains [see Clancey (1983*b*) for further discussion].

An ordered collection of metarules constitutes a procedure for achieving a task. Each metarule has a premise, which indicates when the metarule is applicable, and an action, indicating what should be done whenever the premise is satisfied. Figure 3 is a high-level abstraction of a task and its metarules. The premise looks in the domain knowledge base or the problem-solving history for findings and hypotheses with certain properties, for example, possible follow-up questions for a recent finding or a subtype

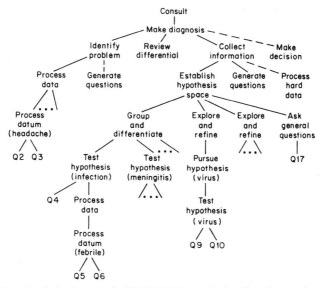

FIG. 2. Invocation of tasks in the example NEOMYCIN consultation. Question numbers correspond to questions asked in the consultation, solids lines show tasks actually done, broken lines those which might be done. Note how such tasks as TEST-HYPOTHESIS are invoked multiple times by a given task as well as by different tasks.

of an active hypothesis. Associated actions would be to ask the user a question or call a task to refine the hypothesis under consideration. The metarules associated with a task may describe the sequence of steps used to achieve the task (in which case the applicable rules are applied once in order), or may present alternate strategies for achieving the goal (in which case the preferentially ordered rules are executed until the goal of the task is achieved).

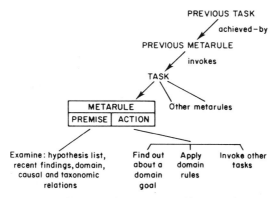

FIG. 3. Abstraction of a task and its metarules.

Testing an hypothesis is just one reason for asking a question. Others are:

initial identification of the problem:
follow-up questions to specify recent information in greater detail;
questions triggered by recent data that suggested an hypothesis;
general questions to determine the completeness of the case history;
screening questions to determine the relevance of desired data; and
questions asked while pursuing subgoals of the domain.

The number of reasons for asking questions testifies to the complexity of NEOMYCIN's diagnostic strategy and illustrates why strategic explanations are both useful and interesting.

4. NEOMYCIN's strategic explanations

4.1. WHY AND HOW EXPLANATIONS

Excerpts from a NEOMYCIN consultation illustrate the strategic explanation capability most clearly. User input follows the double asterisks (**) and is shown in capital letters. The consultation begins:

```
        Please enter information about the patient.
              Name                    Age        Sex        Race
        1) ** MARY                    42          F          LATINO
        Please describe the chief complaints:
        ** HEADACHE
        ** STIFF-NECK
        ** NAUSEA
        **

        2) How many days has Mary's headache lasted?
        ** 10

        3) How severe is Mary's headache (on a scale of 0 to 4
           with 0 for very mild and 4 for very severe)?
        ** WHY
        [i.e. WHY is it important to determine the severity of
            Mary's headache?]

        [1.0] We are trying to characterize the data just
            obtained and develop the differential.
        Headache severity is a charactizing feature of headache.
        [back to question 3 . . .]
```

To explain WHY something is being done by NEOMYCIN, we first restate the question as "WHY are we trying to do this task?". Because each task is invoked by a metarule in another task (see Fig. 3), we answer the query by stating the task and metarule that invoked the task in question. At the lowest level, there are implicit tasks of requesting data and applying rules. The template for an initial WHY explanation is:

```
        ⟨request for data⟩
        * WHY
        [i.e. WHY is it important to determine ⟨data⟩?]
```

[1.0] ⟨immediately preceding task⟩

⟨what is true about the domain knowledge base or the problem-solving history that enables the metarule that accomplishes this task to succeed⟩

In this case the current metarule is:

METARULE073
IF there is a datum that can be requested that is a
    characterizing feature of the recent finding
    that is currently being considered,
THEN find out about the datum.

We mentioned earlier the idea of providing concrete or abstract explanations as appropriate. In the context of a diagnostic consultation, it seems reasonable to provide concrete explanations when possible. Abstract explanations can always be produced, however. The abstract explanation in this situation is:

** ABSTRACT

[i.e. WHY is it important to determine the severity of
    Mary's headache?]

[1.0] We are trying to characterize the data just obtained
    and develop the differential.

A datum that can be requested is a characterizing feature
    of the recent finding that is currently being
    considered.

When subsequent WHYs are given, we ascend the strategy tree, as illustrated below.

** WHY

[i.e. WHY are we trying to characterize the data just
    obtained and develop the differential?]

[2.0] We are trying to give thorough consideration
    to any recent findings.

The fact that Mary has suffered from headaches recently
    is a new finding, which first must be characterized
    and then used to assist the diagnosis.

[back to question 3 . . .]

** WHY

[i.e. WHY are we trying to give thorough consideration to
    any recent findings?]

[3.0] We are trying to gather identifying information--
    the chief complaint and major problems.

The user must state the problem.

[back to question 3 . . .]

** 3

DIFFERENTIAL:
(VIRUS 564) (CHRONIC-MENINGITIS 422)

Questions 4 and 14 below are both asked as a result of the TEST-HYPOTHESIS task. Notice how the explanations of this single task reflect the different situations of the two WHY questions. Different metarules apply—question 4 is based on a strong "triggering" relation, while question 14 is based on only moderate evidence.

> 4) Does Mary have a fever?
> ** WHY
>
> [i.e. WHY is it important to determine whether Mary has
> a fever?]
>
> [4.0] We are trying to decide whether Mary has an
> infection.
> Fever is strongly associated with infection.
>
> [back to question 4 . . .]
> ** YES
>
> .
>
> .
>
> .
>
> 14) Does Mary have a history of granulomas on biopsy of
> the liver, lymph node, or other organ?
> ** WHY
>
> [i.e. WHY is it important to determine whether Mary has a
> history of granulomas on biopsy of the liver, lymph
> node, or other organ?]
>
> [19.0] We are trying to decide whether Mary has tuberculosis.
>
> A history of granulomas on biopsy of the liver, lymph node,
> or other organ makes tuberculosis more likely.
>
> [back to question 14 . . .]

HOW is asked in terms of a previously mentioned task. The explanation presents any metarules associated with the relevant instance of the task that have been completed, as well as the one currently being executed. As a result of the structure of tasks and metarules, the WHY/HOW explanations have an interesting property. A WHY explanation essentially states the premise of the metarule; this is exactly the reason the metarule succeeded. A HOW explanation is a statement of the action of a metarule; this is exactly what was done.

In the next example, we are asking HOW about a low-level task. Note that the answer is at the lowest level of applying rules.

> ** HOW 19.0
>
> [i.e. HOW are we trying to decide whether Mary has
> tuberculosis?]
>
> We did the following:
>
> [20.1] Applied a rule mentioning a causal precursor to
> tuberculosis (tb risk factor; rule525), and
> [20.2] Gave thorough consideration to any recent findings.
>
> [20.3] Applied a rule mentioning general evidence for
> tuberculosis (focal neurological signs; rule366), and
> [20.4] Gave thorough consideration to any recent findings.
>
> [20.5] Applied a rule mentioning general evidence for
> tuberculosis (compromised host; rule309), and
> [20.6] Gave thorough consideration to any recent findings.

We are now:

[20.7] Applying a rule mentioning general evidence for tuberculosis (a history of granulomas on biopsy of the liver, lymph node, or other organ; rule002).

We will next:

[20.8] Give thorough consideration to any recent findings.

[back to question 14 . . .]

** NO

The HOW question below refers to a high level task (ESTABLISH-HYPOTHESIS-SPACE) mentioned in the omitted sequence of WHYs. Notice that the explanation is at a high level and is entirely abstract.

17) Has Mary been hospitalized recently?
** WHY

.
.
.

** HOW 25.0

[i.e. HOW are we trying to develop the differential using basic history and pertinent physical exam?]

We did the following:

[26.1] Got a general idea of the problem: categorized it · into one of several pathogenic classes or disease loci, or both.

[26.2] Confirmed and refined the differential diagnosis through specific questions.

We are now:

[26.3] Rounding out the diagnostic information by looking generally into past medical history and by reviewing systems.

[back to question 17 . . .]

** NO

Besides these strategic WHYs and HOWs, the user can ask about the current hypothesis, the set of hypotheses currently being considered, and evidence for hypotheses at the domain level.

4.2. COMPARISON TO MYCIN

NEOMYCIN uses an explanation approach similar to MYCIN's, that of explaining its actions in terms of goals and rules, so a brief comparison of the two systems is useful (Fig. 4).

The structure of explanations is parallel, except that in MYCIN rules invoke subgoals through their premises, while NEOMYCIN metarules invoke subtasks through their actions. In fact, NEOMYCIN's rules, which are in the format:

If ⟨premise⟩
Then invoke subtasks

MYCIN	NEOMYCIN

Basic reasoning:
  goal → rule → subgoal
	Basic reasoning:
	  task → metarule → subtask

A goal is pursued to satisfy the premise of a domain rule (backward chaining)	A task is pursued when executing the action of a metarule (forward reasoning with rule sets)
To explain *why* a goal is pursued, cite the domain rule that uses it as a subgoal (premise)	To explain *why* a task is done, cite the metarule that invokes it (action)
To explain *how* a goal is determined, cite the rules that conclude it	To explain *how* a task is accomplished, cite the metarules that achieve it

FIG. 4. Comparison of MYCIN and NEOMYCIN explanations.

could be rewritten in the MYCIN style of:

If ⟨premise⟩
and subtasks done
Then higher task achieved.

However, we have no specific conclusion to make about the higher task, so the actions of all metarules for a given task would be identical. Moreover, the subtasks are clearly different from the database look-up operations of the premise. It is therefore natural to view the subtasks as actions. What makes NEOMYCIN's explanations qualitatively different from MYCIN's is that they are generated at the level of general strategies, instantiated with domain knowledge, when possible, to make them concrete.

4.3. INTEGRATING METALEVEL AND BASE-LEVEL GOALS

Our attempts to provide strategic explanations have clarified for us some of the basic differences between metarules and domain rules. Originally, we thought that tasks were logically equivalent to domain goals, as metarules were the analog of domain rules. Specifically, when Neomycin asked a question, we thought that the stack of operations would show a sequence like this:

task 1
metarule 1
task 2
metarule 2
.
.
.
task n
metarule n
domain goal 1
domain rule 1
goal 2
backward ⎧ .
chained ⎨ .
rules ⎩ .
goal m = question asked of the user

Under this goal-rule-goal scheme, WHY questions could proceed smoothly from the domain level to the metalevel. But in fact, metarules sometimes invoke a specific domain rule directly, so the following sequence occurs:

task n
metarule n
domain rule 1
goal 1
.
.
.
goal m = question asked of the user

In this case, there is an implicit task of "apply a domain rule" (invoked by metarule n). Identifying and explaining implicit tasks like this is what we mean by the problem of integrating metalevel and base-level goals. In MYCIN, when Davis (1976) cites the domain rule being applied, he is skipping the immediate intervening metalevel rationale: "We're asking a question to achieve the goal because we were unable to figure out the answer from rules", or "For this goal, we always ask the user before trying rules". In a more recent version of NEOMYCIN, we do make this rational explicit; however, this is uninformative for most users, and the explanation should properly proceed to higher tasks.

4.4. IMPLEMENTATION ISSUES

We mentioned earlier that NEOMYCIN was designed with the intent of guiding a consultation with a general diagnostic strategy. A given task and associated metarules may be applied several times in different contexts in the course of the consultation, for example, testing several hypotheses. To produce concrete explanations, we keep records whenever a task is called or a metarule succeeds; this is sometimes called an *audit trail*. Data such as the focus of the task (e.g. the hypothesis being tested) and the metarule that called it are saved for tasks. Metarules that succeed are linked with any additional variables they manipulate, as well as any information that was obtained as an immediate result of their execution, such as questions that were asked and their answers. When an explanation of any of these is requested, the general translations are instantiated with this historical information.

Figure 5 presents several metarules for the TEST-HYPOTHESIS task translated abstractly. A sample of the audit trail created in the course of a consultation is shown in Fig. 6; this is a snapshot of the TEST-HYPOTHESIS task after question 14 in the

METARULE411
IF The datum in question is strongly associated with the
    current focus
THEN Apply the related list of rules
Trans: ((VAR ASKINGPARM) (DOMAINWORD "triggers") (VAR CURFOCUS))

METARULE566
IF The datum in question makes the current focus more likely
THEN Apply the related list of rules
Trans: ((VAR ASKINGPARM) "makes" (VAR CURFOCUS) "more likely")

FIG. 5. Sample NEOMYCIN metarules for the TEST-HYPOTHESIS task.

TEST-HYPOTHESIS

STATIC PROPERTIES

TRANS: ((VERB decide) whether * has (VAR CURFOCUS))
TASK-TYPE: ITERATIVE
TASKGOAL: EXPLORED
FOCUS: CURFOCUS
LOCALVARS: (RULELST)
CALLED-BY: (METARULE393 METARULE400 METARULE171)
TASK-PARENTS: (GROUP-AND-DIFFERENTIATE PURSUE-HYPOTHESIS)
TASK-CHILDREN: (PROCESS-DATA)
ACHIEVED-BY: (METARULE411 METARULE566 METARULE603)
DO-AFTER: (METARULE332)

AUDIT TRAIL

FOCUS-PARM: (INFECTIOUS-PROCESS MENINGITIS VIRUS
                          CHRONIC-MENINGITIS MYCOBACTERIUM-TB)
CALLER: (METARULE393 METARULE400 METARULE171 METARULE171
                  METARULE171)
HISTORY: [(METARULE411 ((RULELST RULE423)
                                   (QUES 4 FEBRILE PATIENT-1 RULE423)))
                (METARULE411 ((RULELST RULE060)
                                   (QUES 7 CONVULSIONS PATIENT-1
                                      RULE060)))
             :
             :
          (METARULE566 ((RULELST RULE525)
                                   (QUES 11 TBRISK PATIENT-1 RULE525))
                METARULE603
                ((RULELST RULE366)
                  (QUES 12 FOCALSIGNS PATIENT-1 RULE366))
                METARULE603
                ((RULELST RULE309)
                  (QUES 13 COMPROMISED PATIENT-1 RULE309))
                METARULE603
                   ((RULELST RULE002)
                   (QUES 14 GRANULOMA-HX PATIENT-1 RULE002]

FIG. 6. Sample task properties.

consultation excerpt. An example of how the general translations thus relate to the context of the consultation can be seen in the differing explanations for questions 4 and 14, both asked because an hypothesis was being tested.

In order to generate explanations using an appropriate vocabulary for the user, we've identified general words and phrases used in the translations that have parallels in the vocabulary of the domain. At the start of a consultation, the user identifies himself as either a "domain" or "system" expert. Whenever a marked phrase is encountered while explaining the strategy, the corresponding domain phrase is substituted for the medical expert. For example, "triggers" is replaced by "is strongly associated with" for the domain expert.

5. Lessons and future work

The implementation of NEOMYCIN's explanation system has shown us several things. We've found that for a program to articulate general principles, strategies should be represented explicitly and abstractly. They are made explicit by means of a representation in which the control knowledge is explicit, that is, not embedded or implicit in the domain knowledge, such as in rule clause ordering. In NEOMYCIN this is done by using metarules, an approach first suggested by Davis (1976). The strategies are made abstract by making metarules and tasks domain-independent. We've seen that it is possible to direct a consultation using this general problem-solving approach and that resulting explanations are, in fact, able to convey this strategy. As far as the utility of explanations of strategy, trials show that, as one might expect, an understanding of domain level concepts is an important prerequisite to appreciating strategic explanations.

In regard to representation issues, we've found that if control is to be assumed by the tasks and metarules, *all* control must be encoded in this way. Implicit actions in functions or hidden chaining in domain level rules lead to situations which do not fit into the overall task structure and cannot be adequately explained. This discovery recently encouraged us to implement two low-level functions as tasks and metarules, namely MYCIN's functions for acquiring new data and for applying rules. Not only do the resulting explanations reflect more accurately the actual activities of the system, they're also able to convey the purpose behind these actions more clearly.

There is still much that can be done with NEOMYCIN's strategic explanations. We mentioned that our current level of detail includes every task and metarule. We'd like to develop discourse rules for determining a reasonable level of detail for a given user. We also plan to experiment with summarization, identifying the key aspects of a segment of a consultation or the entire session. We might also explain why a metarule failed, why metarules are ordered in a particular way, and the justifications for the metarules. An advantage of our abstract representation of the problem-solving structure is that when the same procedure is applied in different situations, the system is able to recognize this fact. This gives us the capability to produce explanations by analogy, another area for future research.

The design and implementation of the NEOMYCIN explanation system is primarily the work of Diane Warner Hasling, in partial fulfillment of the Master's degree in Artificial Intelligence at Stanford University. We gratefully acknowledge the assistance of Bruce Buchanan, Ted Shortliffe, and Derek Sleeman. Bill Swartout provided us with abstracts of research on explanation presented at the Idylwild Conference in June 1982. This research has been supported in part by ONR and ARI contract N00014-79C-0302 and NR contract 049-479. Computational resources have been provided by the SUMEX-AIM facility (NIH grant RR00785).

References

AIKINS, J. S. (1980). Prototypes and production rules: a knowledge representation for computer consultations. *Ph.D. thesis*, Stanford University (STAN-CS-80-814).

BENBASSAT, J. & SCHIFFMANN, A. (1976). An approach to teaching the introduction to clinical medicine. *Annals of Internal Medicine*, **84**(4), 477–481.

BROWN, J. S. & BURTON, R. R. (1980). Diagnostic models for procedural bugs in basic mathematical skills. *Cognitive Science*, **2**, 155–192.

BROWN, J. S., COLLINS, A. & HARRIS, G. (1978). Artificial Intelligence and learning strategies. In O'NEIL, H., Ed., *Learning Strategies*. New York: Academic Press.

BROWN, J. S., BURTON, R. R. & deKLEER, J. (1982). Pedagogical, natural language and knowledge engineering techniques in SOPHIE I, II, and III. In SLEEMAN, D. & BROWN, J. S., Eds, *Intelligent Tutoring Systems*, pp. 227–282. London: Academic Press.

CHI, M. T. H., FELTOVICH, P. J. & GLASER, R. (1981). Categorization and representation of physics problems by experts and novices. *Cognitive Science*, **5**, 121–152.

CLANCEY, W. J. (1979). Transfer of rule-based expertise through a tutorial dialogue. *Ph.D. thesis*, Stanford University (August) (STAN-CS-769).

CLANCEY, W. J. (1981). Methodology for building an intelligent tutoring system. *Technical Report*, Stanford University, 1981 (STAN-CS-81-894, HPP-81-18). (Also to appear in KINTSCH, W., POLSON, P. & MILLER, J., Eds, *Methods and Tactics in Cognitive Science*. Hillsdale, New Jersey: Lawrence Erlbaum Associates.)

CLANCEY, W. J. (1983a). The epistemology of a rule-based expert system: a framework for explanation. *Artificial Intelligence*, **20**(3), 215–251.

CLANCEY, W. J. (1983b). The advantages of abstract control knowledge in expert system design. In *Proceedings of AAAI-83*, pp. 74–78.

CLANCEY, W. J. & LETSINGER, R. (1981). NEOMYCIN: reconfiguring a rule-based expert system for application to teaching. In *Proceedings of the Seventh IJCAI*, pp. 829–836.

DAVIS, R. (1976). Applications of meta-level knowledge to the construction, maintenance and use of large knowledge bases. *Ph.D. thesis*, Stanford University (July) (STAN-CS-76-552, HPP-76-7).

DAVIS, R. (1980). Meta rules: reasoning about control. *Artificial Intelligence*, **15**, 179–222.

deKLEER, J. & BROWN, J. S. (1982). Assumptions and ambiguities in mechanistic mental models. In GENTNER, D. & STEVENS, A. S., Eds, *Mental Models*. Lawrence Erlbaum Associates, Inc.

ELSTEIN, A. S., SHULMAN, L. S. & SPRAFKA, S. A. (1978). *Medical Problem Solving: An Analysis of Clinical Reasoning*. Cambridge, Massachusetts: Harvard University Press.

GENESERETH, M. R. (1982). The role of plans in intelligent teaching systems. In SLEEMAN, D. & BROWN, J., Eds, *Intelligent Tutoring Systems*, pp. 136–156. London: Academic Press.

LANGLOTZ, C. & SHORTLIFFE, E. H. (1983). Adapting a consultation system to critique user plans. *Technical Report*, Stanford University (April) (HPP-83-2).

MANN, W. C., BATES, M., GROSZ, B., McDONALD, D., McKEOWN, K. & SWARTOUT, W. (1981). Text generation: the state of the art and the literature. *Technical Report RR-81-101*, ISI (December).

McKEOWN, K. R. (1982). Generating natural language text in response to questions about database structure. *Ph.D. thesis*, University of Pennsylvania. Published by University of Pennsylvania as *Technical Report MS-CIS-82-5*.

PATIL, R. S., SZOLOVITS, P. & SCHWARTZ, W. B. (1981). Causal understanding of patient illness in medical diagnosis. In *Proceedings of the Seventh IJCAI*, pp. 893–899 (August). (Also to appear in CLANCEY, W. J. & SHORTLIFFE, E. H., Eds, *Readings in Medical Artificial Intelligence: The First Decade*. New York: Addison–Wesley.)

POPLE, H. (1982). Heuristic methods for imposing structure on ill-structured problems: the structuring of medical diagnosis. In SZOLOVITS, P., Ed., *Artificial Intelligence in Medicine*. Boulder, Colorado: Westview Press.

SCOTT, A. C., CLANCEY, W., DAVIS, R. & SHORTLIFFE, E. H. (1977). Explanation capabilities of knowledge-based production systems. *American Journal of Computational Linguistics*, microfiche 62.

SHORTLIFFE, E. H. (1976). *Computer-based Medical Consultations: MYCIN*. New York: Elsevier.

SWARTOUT, W. R. (1981a). Producing explanations and justifications of expert consulting programs. *Ph.D. thesis*, Massachusetts Institute of Technology (January) (MIT/LCS/TR-251).

SWARTOUT, W. R. (1981b). Explaining and justifying expert consulting programs. In *Proceedings of the Seventh IJCAI*, pp. 815–822 (August).

WALLIS, J. W. & SHORTLIFFE, E. H. (1982). Explanatory power for medical expert systems: studies in the representation of causal relationships for clinical consultations. *Methods of Information in Medicine*, **21**, 127–136.

WEINER, J. (1980). BLAH, a system which explains its reasoning. *Artificial Intelligence*, **15**, 19–48.

WILLIAMS, M., HOLLAN, J. & STEVENS, A. (1981). An overview of STEAMER: an advanced computer-assisted instruction system for propulsion engineering. *Behavior Research Methods & Instrumentation*, **13**, 85–90.
WINOGRAD, T. (1972). *Understanding Natural Language*. New York: Academic Press.

Expert systems: an alternative paradigm

MIKE COOMBS AND JIM ALTY

Department of Computer Science, University of Strathclyde, Glasgow G1 1XH, Scotland, U.K.

There has recently been a significant effort by the A.I. community to interest industry in the potential of expert systems. However, this has resulted in far fewer substantial applications projects than might be expected. This article argues that this is because human experts are rarely required to perform the role that computer-based experts are programmed to adopt. Instead of being called in to answer well-defined problems, they are more often asked to assist other experts to extend and refine their understanding of a problem area at the junction of their two domains of knowledge. This more properly involves educational rather than problem-solving skills.

An alternative approach to expert system design is proposed based upon guided discovery learning. The user is provided with a supportive environment for a particular class of problem, the system predominantly acting as an advisor rather than directing the interaction. The environment includes a database of domain knowledge, a set of procedures for its application to a concrete problem, and an intelligent machine-based advisor to judge the user's effectiveness and advise on strategy. The procedures focus upon the use of user generated "explanations" both to promote the application of domain knowledge and to expose understanding difficulties. Simple database PROLOG is being used as the subject material for the prototype system which is known as MINDPAD.

1. Experts and expertise

Over recent years there has been a significant effort by the A.I. community on both sides of the Atlantic to awaken industry to the potential of knowledge-based systems. As a result, many large companies have set up working groups, industry is well represented at A.I. conferences, and universities have run well-attended expert systems tutorials. There is ample evidence of interest. However, all this activity has resulted in far fewer substantial applications projects than might be expected.

The authors have been recently involved in setting up two expert system projects within the very different application areas of paint selection for industrial machinery and computer-user support. In both of these projects, preparatory investigations have revealed the same basic difficulty in applying expert systems technology, namely that human experts are rarely called upon to perform the role that current computer-based experts are programmed to adopt. Rather than acting as "wise-men" called upon to produce solutions to complex—but essentially well-defined—problems, human experts in these domains are more often asked to provide conceptual guidance to other experts in adjacent fields to enable them to solve problems for themselves.

Instead of being valued as diagnosticians, experts in the above domains are employed for their ability to assist colleagues to extend and refine their understanding of the problem area at the junction of the two fields of knowledge. In the case of the paint selection system, for example, this might involve the relationship between a food processing expert's knowledge of the toxicity of certain paint constituents and a paint

DEVELOPMENTS IN EXPERT SYSTEMS
ISBN 0-12-187580-6

expert's knowledge of the adhesive properties of such constituents. The promotion of understanding is achieved by such activities as:

(a) providing relevant contextual information;
(b) focusing attention on important topics in the subject area;
(c) helping to predict outcomes of given processing circumstances.

Many of these functions involve educational rather than traditional problem-solving skills (for example, Newell & Simon, 1972) and their support appears to be conceptually outside the range of current expert systems.

Expert systems are designed to achieve known and clearly defined solutions to a well-circumscribed class problem. This is true both of systems which encode the loosely structured, empirical knowledge of experts [for example, MYCIN (Shortliffe, 1976) and DENDRAL (Feigenbaum, Buchanan & Lederberg, 1971)], and systems which reason from some representation of the "causal" relationships underlying the problem domain [for example, CASNET (Kulikowski & Weiss, 1982) and Davis (1983)]. The above programs are all clearly problem-solvers in that their sole objective is to achieve known and clearly defined solutions to a well-circumscribed class of problem. MYCIN's purpose is to diagnose and recommend treatments for bacterial infections of the blood; DENDRAL analyses and evaluates spectographs and selects the most likely composition of substances. Moreover, as problem-solvers, their goals remain the same each time they are used (i.e. to recommend treatments, to name substances).

As we argue in detail below, the critical difference between the role of an expert as a problem-solver and as an advisor is that, while the former focuses upon the process of obtaining a concrete, communicable solution to a problem, the latter is primarily concerned with the enrichment of the user's understanding of a problem area and the development of his skills at handling that area. As an advisor, the expert is expected to support a colleague's personal problem-solving, particularly at the junction between their two areas of expertise, and to help him decide what questions should be asked and how to look for answers.

Implemented as a knowledge-based computer system, a guidance program for the areas we have studied would need to reverse the priorities of a conventional diagnostic expert system, focusing upon the elaboration of content relevant to problem-solving rather than on the solution itself. The knowledge-base would need to be highly detailed, containing many different types of information (theoretical, empirical, pragmatic), and open to changes throughout a guidance session. It must also be possible for the knowledge to be applied to novel goals generated by the user, enabling him to obtain answers to a wide range of different types of question, including:

what would happen if X?
• why would X happen?
how could X be prevented?
what are the critical factors in X?

Formal approaches to answering such questions have been made in the development of A.I. text-understanding systems (Lehnert, 1978), but have yet to be developed in a guidance context.

A number of attempts have been made to extend diagnostic agents to enable them to support an advisory role. However, such efforts have revealed a number of substantial difficulties. A principal one of these concerns the content and structure of the knowledge-base, and is illustrated persuasively by the GUIDON project (Clancey, 1979). Following the success of MYCIN as a diagnostic program, it was decided to develop it into a tutorial system using the original knowledge-base in the new role of guiding and evaluating a student's learning. However, it was found that much of the knowledge required by a person attempting to learn how to problem-solve within the subject area (as distinct from using the system to support his problem-solving) proved to be implicit within the diagnostic rules, and so not available for inspection. It thus proved necessary to augment substantially the knowledge-base, including new information about subject primitives, the structure of concepts and suitable learning strategies [see NEOMYCIN (Clancey & Letsinger, 1981)].

A second source of difficulty concerns system architectures. The TEIRESIAS interface to MYCIN (Davis & Lenat, 1982), for example, was intended to provide an adaptable explanation system to help both user and subject expert to understand the system's reasoning. However, much of the power of TEIRESIAS depended upon the highly structured nature of content and the use of meta-rules for control. It might be suggested that the need for a highly structured subject domain places conceptual limits on the approach, and the extensive use of meta-rules could eventually be self-defeating by making the implications of conceptual relations within the system opaque. This is obviously not desirable in a system with the express purpose of developing understanding.

Early in 1982, the authors decided to explore possible designs for an intelligent computer-based guidance system. The system—MINDPAD—would be explicitly concerned with supporting the conceptual aspects of a user's efforts at problem-solving; providing an environment in which the user may achieve a better understanding of his problem, and in so doing be equipped to solve it. The basic requirements for this, and some principles for its provision, emerged during a study of interactive problem-solving between university computer users and professional advisors undertaken by the authors and discussed below (Alty & Coombs, 1980, 1981; Coombs & Alty, 1980, 1982).

2. Expert guidance: results of the Advisory Service Study

The Advisory Service Study was set up with the explicit objective of improving the effectiveness of computing advisory services in university computer centres. These services have been established in most centres to provide some measure of personal support in computing. This is necessary because of the very wide range of expertise found in any population of university users and the desire of many researchers to tap the power of the computer with as little investment in fundamental training as possible. Any realistic support for such people must be able to adapt to the needs of individuals, a goal which has proven difficult to achieve with conventional documents and automated "help" facilities (Alty & Coombs, 1980).

In order to study interactions, random samples of conversations were recorded at five university computer centres. Soon after recording the participants were debriefed in depth, with reference to both the conversation text and the relevant documentary material. The debriefing was goal based, focusing upon such factors as the general motivation(s) for the session, the goal(s) behind each individual utterance, and the success with which goals were achieved (Coombs & Alty, 1980).

Interactions took place at the user's initiative and without appointment. They were mainly concerned with the diagnosis and correction of failure in some item of software, this frequently involving in some way the user's own program. The success rate for both diagnosis and correction was high (around 80%), and was normally achieved on the first visit during an interaction lasting not more than 10 minutes.

Viewed in retrospect, a striking feature of the majority of advisory conversations was the extent to which they resembled the style of interaction supported by rule-based systems. Given that such interactions were often considered unsatisfactory by users in spite of a high success rate, and given the contrast of a few interactions with a different style which were considered most satisfactory, it was decided to review the advisory study data in more detail for guidance on the design of our system.

The interactions considered as unsatisfactory proved to be most strongly controlled by the advisor, the user only being required to supply information where it could not be obtained elsewhere. There was also little feedback to the user after questions and no explicit account of how the information was to be used for reasoning. In general, all reasoning was covert, unless some explanation was explicitly requested. Advisors made no attempt to educate users. Conversations lacked any elaboration of the facts or terms employed by advisors (e.g. explanation of the error message "OVERFLOW" in terms of the methods used to store numbers in computers), contained no review of the methods used to obtain the successful diagnosis, and no review of the solution.

The effect of the above style emerged clearly during debriefing. First, the problem solved was sometimes a restricted (often the most concrete) portion of what was a much broader problem, with many conceptual ramifications. As might be expected, this was most frequently the case with inexperienced users who did not have the knowledge to verbalize the difficulty they "felt" themselves to have, and so resorted to asking something close and within their conceptual scope. The least well-prepared users were thus left to themselves both to define their real problem and to generalize the solution to it. Secondly, users were often left in doubt as to the correctness of their response to a question, given that advisors often did not appear to use the information provided. Because of this, these questions proved less useful as anchors to aid in their learning of successful reasoning than might have been expected. Thirdly, advisors failed to elaborate the context within which the advice could be understood, simply giving a bold assertion of the problem and the steps necessary for its solution. This made both recall of the solution and the reconstruction of reasoning very difficult.

As mentioned above, a small number of interactions were characterized by a very different style and did not suffer from these difficulties. It is features of these sessions that we have incorporated into the design of our computer-based guidance system. They usually took place between a fairly-experienced user and a local computer expert who had a passing knowledge of the problem domain but no specific knowledge about the particular problem in hand. The two participants thus shared the roles of advisor and client, each within their respective area of expertise. However, this resulted in conversations which were notable for their apparent lack of structure, both participants making contributions in a loose and uneconomic manner. Often, much more information was given than appeared to be strictly relevant. Nevertheless, analysis of the goals of participants collected at debriefing indicated that this class of interaction was well-motivated despite surface appearances, the objective not being strict problem-solving as we had assumed, but problem-solving through mutual understanding. This required sensitivity to different structural factors.

The most important factor in this second group of interactions was the extent to which both advisors and users made public and explicit much of what was covert and implicit with the majority of sessions. This had two effects. First, it was possible for participants to monitor each other throughout the interaction, and to provide effective feedback. Secondly, the public nature of the session had the side-effect of forcing a greater degree of precision in all aspects of the interaction and of forcing participants into a "deeper" understanding of their mutual information and inferential needs during the solution of the problem.

With reference to the improved feedback, this proved to help participants to develop a better understanding of the process of solution. With reference to the public nature of the session, there were many additional advantages. However, most importantly these included the "focusing" of relevant domain facts onto the problem and the clarification of processing goals (the two participants often stepping outside each other's problem view to suggest alternative, and perhaps more productive, approaches). Finally, the very act of participation appeared to help develop a better shared understanding of the problem and its structure. Both participants were often able to gain more fully from the session by abstracting computing facts and problem-solution methods of some general value.

In order to make the transformation from the successful advisory sessions to a computer-based guidance system, it is necessary to isolate the fundamental cognitive procedures underlying interactions. To do this systematically requires some theory of the role of conceptualization and understanding in problem-solving, which will be approached in the next section. However, it is possible to make some headway by asking three questions of the conversations.

(a) Why does the user need to seek help in the first place?
(b) Is there an optimum interaction for a given class of problem?
(c) How may a user's problem-solving skills be increased as a by-product of interactions?

With regard to user motivation, it was found that problems were more often conceptual than procedural; the user lacked, and failed to develop while working alone, adequate concepts for both defining problem goals and for employing them towards a solution. While we do not claim that inadequate procedures were never at fault, problems were often solvable employing skills the user already possessed with a richer set of concepts. Moreover, it was observed that users frequently lacked the ability to develop appropriate concepts in the problem area because they were unaware of what the concepts would look like. For example, a user of the PROLOG language was unable to develop the concept of "backtracking" because he saw variable instantiation as a permanent act.

Given that all new concepts may be argued to be derived from existing concepts (see section 4), this points to some deficiency at a meta-skill level (i.e. procedures for forming concepts appropriate to a given class of problem and problem domain). Difficulties in this area may include the use of an unsuitable "language" for representing the problem (e.g. it is better to think of the process of tree-passing in diagrammatic rather than verbal terms), the overgeneralization of a concept that has failed to achieve a solution (the new concept may allow the problem to be "solved", but in a manner that is not sufficiently concrete to be useful), and over-reliance on an over-elaborated problem context (e.g. using assumed and unverified characteristics of the underlying machine in debugging).

Consideration of the optimum interactive relationship between advisor and user has suggested that as much of the processing as possible should be made explicit. We would also like to add that even with the most successful interactions, difficulties were traced to the limited capacity of human working memory, a participant often being forced to make the same deduction on several different occasions or to rehearse an argument to aid recall. A system for making explicit both concepts and arguments would help in this area.

"Learning to learn" conceptualization and inferential skills is an important outcome of an advisory interaction. This appeared to be achieved best when the user was engaged in the solution of his own problem with the co-operation of the advisor. In education this would be classed as a type of guided discovery learning, with the advisor helping to identify solvable sub-problems, forcing the user to be explicit in describing and justifying his moves, supplying information or methods where required and monitoring progress.

The general problem-solving strategy favoured by participants in the successful interactions involved the generation, and then critiquing, of explanations for some set of problem phenomena. Both generation and critiquing was conducted by both advisor and user, although one participant usually tended to lead more frequently. The focus of this activity was primarily upon individual results and units of explanations usually amounting to relations between two or three assertions. The direction of problem-solving activity was thus primarily conducted bottom-up with attention to the truth of individual assertions and their sequential validity. However, there were often occasions when this strategy would be radically changed, a whole set of errors or unexpected phenomena being asserted as arising from a single act; often this was itself asserted with confidence as following from some conceptual difficulty. This then appeared to trigger a radical switch in approach to a top-down search for confirming evidence (disconfirming evidence occasionally being ignored, or put aside to explain later).

It was this emphasis on a conservative, bottom-up analysis of the problem that often made it difficult to identify the structure in conversations from study of the utterances alone. The actual selection of items of program or explanation to analyse was governed by the participants' models of the subject domain. The setting of work goals was thus achieved top-down, while the drift of problem-solving activity was primarily bottom-up. The process of goal selection, particularly global considerations, sometimes only became clear to the researcher during debriefing, although it was understood by the participants. Moreover, given that both participants had different levels of knowledge concerning the two subject domains, they engaged in a great deal of mutual justification of individual choices of work area, which itself generated problem-solving activity which was difficult to distinguish from the primary activity of solving the user's problem. Finally, there was the additional complication that any interaction involved both the diagnosis of the cause of the current problem and the generation of principles for correction. These two objectives were usually intermixed, particularly at the stage when the participants were near to identifying the problem and were looking ahead to correction.

Following the findings summarized above, it was proposed that our guidance system should support rather than direct problem-solving. This support should mainly be in the form of assistance in building and evaluating concepts in the problem area. The model of support should be that of guided discovery, control being invested equally in the user and the system. The program may be seen as providing a set of tuned resources, some of them passive (e.g. a concept specification language), and some of them active (e.g. a guidance system for assisting the user with problem-solving strategy), to promote conceptual development and application.

3. Theoretical foundations

A clear and fundamental feature of the guidance advisory interactions described above is the dynamic, constructive nature of the information processing undertaken by the user. Within British psychology there has been a long tradition of viewing human cognition in this way, the seminal work being undertaken by Bartlett concerning the nature of memory (Bartlett, 1932). In contrast to the dominant Behaviourist psychological philosophy of the period, Bartlett argued that the results of work on memory for complex, meaningful material is best explained in terms of active information processing. Learning is seen as a process of building models, or "schemata" of the subject material which is then used for its reconstruction on recall. The "schemata" may be interpreted as procedures which act to control remembering, and as a focus for the "attachment" of supporting procedures where the main procedures fail.

This approach is present in the work of many others on both sides of the Atlantic and in a range of different application areas, including psychiatry (Kelly, 1955), education (Bruner, 1957; Piaget, 1972) and visual perception (Gregory, 1970). However, while theoretically promising, much of the work failed to develop into a general model of cognition that was sufficiently powerful for application purposes. The deficiency usually lay in the absence of both the conceptual and methodological tools to represent mental procedures and their transformation. There has been much recent progress in this direction, being principally centred on A.I. [for example, Schank's work on the use of Scripts (Schank & Abelson, 1977) and later MOPs (Schank, 1979) in text understanding]. However, these models are still not adequate for application purposes, lacking sufficient predictive power. The objective of much A.I. now is to define sufficient structure and content to achieve given processing goals, rather than the structure and content necessary for their attainment. They thus lack the appropriate formal foundation for the construction of powerful theories [the work of Lenat (Lenat & Harris, 1978) is one exception]; there is no systematic investigation into the necessary rules of macrostructure such as we find in linguistics (for example, van Dijk, 1980).

Out of the many disciplines concerned with information processing that have developed this century, cybernetics is notable for being concerned with questions of necessity. Its fundamental concern is to establish the minimal theoretical structure required for some given class of processing. Within cybernetics, the most comprehensive application of this approach to issues of learning and understanding is found within the work of Pask (1975, 1976), which aims to produce a general theory of cognition rather than solely to explain specific cognitive episodes. This aspect of Pask's work has proved attractive for establishing a general framework within which the specific requirements of a particular problem-solving type and subject area may be modelled.

Pask (1975) proposes that the minimum structure for learning requires at least two processors: one to interact directly with the world, and one to modify the procedures that undertake such interaction.

A diagrammatic representation of this structure is given in Fig. 1, which follows that presented by Pask (1975). Each of the boxes represents some domain, the arrow across a box indicating that the procedures within it do something to the domain. The symbol "*" and the arrow deriving from it indicates a feedback path between domains. The domain interacting directly with the world is labelled "Level 0" and contains "p0" procedures; the procedure modifying domain is labelled "Level 1" and contains meta-procedures "p1". This structure can be seen as representing a program like HACKER (Sussman, 1975) which diagnoses faults in its procedures and corrects them accordingly.

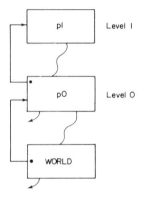

FIG. 1. A representation of the minimum structure for learning (after Pask, 1975).

Examination of Fig. 1 will reveal that the structure has a limitation not possessed by humans: learning will only take place within the limits of the procedures stored at Level 1. However, humans have the ability to learn new ways of learning—learning strategies. To do this, the system will need to be able to re-build, or re-interpret, itself in some novel manner. The minimum structure proposed by Pask (1975) for doing this is given in Fig. 2.

In order to re-build or re-interpret its procedures, the system must have some account of the former structures. There are thus now three units at each level (conceptually, at least): the newly built procedures (p′); a description of the old procedures (D); the procedure building procedures (p″). Given the two conceptual levels of operation, Pask now proposes the need for two complete but communicating systems. Moreover, these may be seen as residing either within one physical machine (or organism), or distributed between two machines (or organisms). The relationship between the two systems is termed a "Conversation", hence the choice of the term "Conversation Theory" for the whole approach.

Using the Conversational structure given in Fig. 2, Pask goes on to explore a variety of cognitive processes such as learning, problem-solving and thinking. Central to Pask's exposition are the notions of a "concept", "memory", "understanding" and "explanation", which are re-defined within the notion of a Conversation. A Conversation is

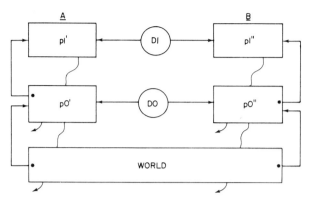

FIG. 2. A Conversational structure (after Pask, 1975).

seen as an interaction between two processors A and B concerning some "topic". The "topic" is defined as a set of connected assertions, each containing an interpreted formal relationship (e.g. "next", "adjacent", "sum"), the interpretation taking place within some context (e.g. statistics, computing, poetry). Examples of "topic"'s would include physical laws, social theories and musical form. Topics are seen as related by the two rules of "subsumption" and "analogy" to form subject domains, which Pask represents as a connected graph of topics organized into topic hierarchies (following the rule of subsumption). In this manner, Pask is in accord with thinking on the structure of a thesaurus, which favours a lattice (Parker-Rhodes, 1978). Pask terms such graphs "entailment nets".

The cognitive terms listed above (e.g. concept, memory, understanding, explanation) may be described as operations upon the topics in an entailment structure. A concept, for example, may be seen as a procedure which brings about (or "satisfies") the relation(s) which is (are) a topic: a memory for a topic may be seen as a procedure which reconstructs or reproduces a concept. The notion of understanding is more complex, it being seen as a product of meta-procedures (at Level 1) which build concepts a (at Level 0), and as evidenced by the ability of an organism (or machine) to provide an explanation of "Why" building proceeds in a particular manner. Evidence for the possession of a concept is simply the "How" description if the activity necessary to bring about the topic relation(s)—at this level, explicit motivation for the steps in the procedure would not be required. It should be noted that all this activity takes place in a co-operative Conversational environment.

4. "MINDPAD"—a prototype guidance system

4.1. SYSTEM OVERVIEW

The view of Guidance we have adopted for MINDPAD derives both from the results of the empirical work done in the Advisory Service study and from the theoretical work of Conversation Theory. Together they are required to provide a conceptual framework for the design of a system which will support a rich variety of human–computer exchange. This section gives a overview of the basic facilities required of such a system.

Interactions with the Guidance program MINDPAD are structured in the context of Conversation Theory to enable the user to develop problem-solving skills with some subject domain by achieving an "understanding" of topics within the domain. The problem-solving skills are thus developed as a direct product of achieving "understanding". This is congruent with the view developed in sections 1 and 2 that the objective of a guidance session [as distinct from a problem-solving session, or indeed a tutorial (Coombs & Hughes, 1982)] is to give the user a measure of creative independence within the subject area. The session should aim to help the user learn at a meta-conceptual level as well as at the base conceptual level. He should, for example, leave the session with the ability to describe the principal relations between topics, to give an account of appropriate problem-solving heuristics, and explain their relevance.

A range of different, but equally valid, guidance procedures may be defined within a Conversational context, and it is intended that such variations will be a subject of our research. However, for the initial prototype system, it seems wise to keep as close as possible to the learning environments employed by Pask to test his theory (Pask, 1976). These may be described broadly as implementations of "guided discovery learning" and have proved to be very effective within an instructional setting.

The system aims to develop the user's understanding of the subject area as a result of his working at a concrete problem. The problem acts as a focus for his efforts to identify and explore concepts and topics within the domain (represented in a frame-like database), his understanding being developed by the discipline of application and by the requirement that all solutions be "explained". Following Pask (1976), explanation is employed as the principal mechanism for generating the descriptions which enable intelligent interaction to take place both between user and system and within the user's own head.

The task area selected for the prototype system is that of debugging simple database PROLOG programs. PROLOG provides a suitable domain for two reasons. First, novices often have difficulty in establishing a stable mental representation of a program because the language has both a procedural and a declarative semantics. This frequently leads to contradictions between application and processing descriptions of programs, the former referencing the programmer's knowledge of his application domain and the latter referencing the details of PROLOG execution. Reasoning about a particular program thus closely resembles the interaction between co-operating experts discussed in the previous sections. Secondly, the recursive and backtracking features of PROLOG can make flow of control difficult to follow and so justify the need for computer guidance. Both the PROLOG and guidance systems are being programmed in FRANZ LISP.

The global structure of the prototype system is given in Fig. 3. It may be seen that the user has access to a number of different resources. Two of these may be described as passive and include:

(a) the STRUCTURE GRAPH, which functions as the knowledge-base for the system and incorporates a PROLOG interpreter;

(b) the WORKPAD, which has the primary function of providing the resources to enable the user to develop problem solutions and to construct explanations. However, it also serves to interface the user to both the STRUCTURE GRAPH and to an intelligent, machine-based ADVISOR.

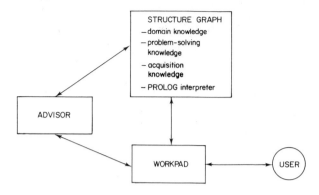

FIG. 3. MINDPAD overview.

The ADVISOR is an *active* resource and has three functions: evaluating user explanations, identifying failures in understanding and guiding users on problem-solving strategy. To permit such activity, the ADVISOR employs problem-solving knowledge stored within the STRUCTURE GRAPH, and has full access to WORKPAD. The problem-solving knowledge concerns methods for identifying explanatory errors and ambiguities, making decisions about user understanding and advising on corrective action. In addition, the ADVISOR runs the current Conversational model and co-ordinates interaction with the various learning resources.

Within the framework of the Conversation diagram given in Fig. 2, the user may be seen as occupying system B and the ADVISOR as occupying system A. The ADVISOR, as mentioned above, has the role of making available resources, critiquing explanations, and suggesting tasks. Advice may be given gratuitously, or may be requested by the user, and the user may accept or reject it as he wishes.

4.2. THE "STRUCTURE GRAPH"

The STRUCTURE GRAPH is the main source of factual knowledge for the system, including descriptions of PROLOG structures, execution processes and comprehension difficulties.

Information is organized into a network of frame-like schema termed "topics". These are related using the single rule of "subsumption". To this extent the STRUC-TURE GRAPH is a simplification of an "entailment net" as defined by Pask (1975), any given topic being interpreted as a complex relation resulting from some operation being applied to the topics falling below it.

Knowledge about PROLOG, for example, may be represented under four groups of related topics: syntax, fundamental structural units, program structure and program execution. A sample PROLOG "sub-graph" is given in Fig. 4. It should be noted that the sub-graphs are not fully independent but are related to each other at nodes marked with angle brackets. Furthermore, the actual topics selected and their organization is not in any sense intended to be necessary. Some other author might have provided an alternative decomposition of the subject matter or some alternative organization for the current decomposition. However, it is intended that the view presented by the

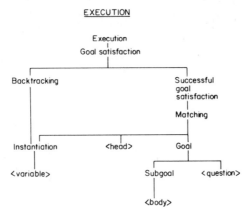

FIG. 4. Topic structure for PROLOG execution.

STRUCTURE GRAPH should be both coherent and sufficient for the problem-solving tasks to be undertaken using the system.

The STRUCTURE GRAPH includes several different classes of information.

(a) PROLOG facts—describing the structure of PROLOG objects and processes of PROLOG execution.

(b) Problem-solving facts—describing the relationships between faults in a user explanations of program execution and comprehension errors.

(c) Facts about knowledge acquisition—descriptions of methods for developing an understanding of PROLOG concepts.

Individual topics are thought of as frame-like entities, their structure being expressed in terms of classes of slot and meta-information on the values that a slot may adopt. Typical slot types include those for the description of substantive domain facts (STRUCTURE, EVENT), those for indexing the topic structure (SUBSUME), and those for recording details of the current interaction. In the case of slots that describe substantive information about the domain, the meta-information is in a form which both allows it to be output as coherent text and which allows it to be instantiated with instances from the current explanation and problem program. A sample topic description for the PROLOG topic of "MATCH" is given in Fig. 5.

It will be noted that the actual descriptions of "MATCH" are given in PROLOG code. This is because a feature of the system is that descriptions should be executable as programs. It has been found that this provides a powerful procedure for constructing and evaluating hypotheses concerning comprehension errors.

Finally, the STRUCTURE GRAPH provides both instructional and reference information to the user. The prototype system will provide a menu of question types that may be addressed to the knowledge-base, concerning both the relationships between topics and the substantive information within topics. Queries at the topic level include:

(a) what topics exist in the domain?
(b) what topics do I need to understand so I can understand topic X?
(c) what topics subsume topic X?

```
*************************************************************
     TOPIC: MATCH

     SUBSUME: INSTANTIATION, HEAD, GOAL

     EVENT: match(head([],goal([|)).
            match(head([H|T1]),goal([H|T2])):-
                  mrule(const(H,H);var(H,H);var(inst(H,H))),
                  match(head(T1),goal(T2)).

     EXAMPLE:
            succeed(match(head([a,b,c]),goal([a,b,c]))).
            succeed(match(head([a,X,c]),goal([a,Y,c]))).
            succeed(match(head([a,b,X]),goal([Y,b,c]))).
            fail(match(head([a,c,d]),goal([a,b,d]))).

     ACCESS: <                                              >

*************************************************************
```

FIG. 5. Details of the topic "MATCH".

Queries at the substantive level include:

(a) tell me about topic X?
(b) is information item X correct?
(c) tell me about information item X?
(d) is information item X correct?
(e) what would happen if X (or not X)?
(f) could X cause (result in) Y?
(g) how is X related to Y?

Procedures for implementing these queries exploit the constraints present in the subsumptive relations between topics and relations permitted within the explanation grammar. They also draw heavily upon the techniques developed by Lehnert (1978).

4.3. "WORKPAD" AND THE EXPLANATION CYCLE

Consultation sessions develop through a cycle of problem specification, solution and explanation which may be mapped onto the Conversational model given in Fig. 2. The complete cycle of revision in program development is given in Fig. 6, although only the part specified by the broken line will be encompassed by the prototype. This focusses upon obtaining a correct explanation of execution of a program, including the execution of a faulted program.

All of this problem-solving activity is mediated through WORKPAD. The program must therefore support a range of different types of interaction. However most of them will be focussed upon the processes of building and critiquing explanations. WORKPAD functions include the following.

(a) Support of user activity:
    construction of solution programs;
    running of solution programs;
    construction of explanations;
    access to information in the STRUCTURE GRAPH;
    access to explanation grammar and dictionary of PROLOG terms.

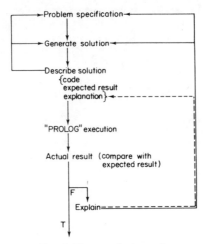

FIG. 6. The consultation cycle.

(b) Support of ADVISOR activity:
   critiquing of explanations;
   presentation of critiques;
   presentation of strategy advice.

At this point it would simplify the exposition to give a summary of the MINDPAD problem-solving cycle. This is intended to provide a context for understanding the operation of WORKPAD, and later the ADVISOR, and so will avoid detailed accounts of these facilities which will be discussed later. Furthermore, the cycle will be described in a form which emphasizes user control and the employment of passive resources for the analysis of user explanations. Although it is anticipated that the system will eventually be able to provide a range of different levels of guidance intervention, the details of appropriate ADVISOR support are as yet not sufficiently clear.

A consultation starts with the user presenting his problem to the system. Throughout the interaction, information concerning the current solution is stored in WORKPAD within a pair of declaration frames: one for the PROLOG code itself and one for an explanation of the code in terms of its execution. The explanation will also give the expected output of the code. The declaration frames are structured to accommodate a particular model of PROLOG (outlined below), so helping to shape the user's concept development and problem-solving.

Following the setting of the initial declaration frames, the user may either run the program that represents his proposed solution, or request a critique of his explanation. If he chooses to run the program, the actual results will be recorded in WORKPAD below the declaration of the expected results. A comparison is then made and any deviation interpreted as indicating a failed solution.

In the event of failure, the user is required to work at understanding the results by analysing the explanation and amending it in order to justify them. This may be done with various levels of system support. These include:

the simple retrieval of PROLOG descriptions from the STRUCTURE GRAPH;
a request for a critique of the current explanation to be given by the ADVISOR;
a consultation with the ADVISOR on problem-solving strategy.

At any point the user may decide that he understands the problem sufficiently to attempt another solution. In this case, he may abandon his effort to justify the previous results and work at composing the solution, along with its related explanation. This may then be run and, if it fails, the cycle is repeated. Before changing activity, the user has the option of saving the program and explanation for later inspection.

WORKPAD implements a "two level" approach to understanding PROLOG programs proposed by Bryd (Bryd, 1980) and used in the design of debugging tools for the DEC-10 PROLOG system (Clocksin & Mellish, 1981). PROLOG programs consist of sets of clauses, each of them beginning with a particular relation name (a predicate). There are two types of clause: facts, which assert a relation with regard to one or more objects [e.g. "likes(mary,wine)." states "Mary likes wine"]; rules, which assert that some set of relations are true, given the truth of some other set of relations [e.g. "likes(john,X):—human(X),likes(X,wine)." states "John likes people who like wine."]. Given these two clauses, and an additional one stating that "Mary is human" (see Fig. 7), it is possible to use the PROLOG interpreter to deduce that "John likes Mary" by getting it to instantiate the variable "X" in the rule with the value "mary".

```
likes(mary,wine).
likes(john,X):-human(X),likes(X,wine).
human(mary).
```

FIG. 7. A simple PROLOG program.

PROLOG undertakes the deduction by taking the question "Who does John like?" in the form "?-likes(john,X1)." and seeking to match it to a clause within the program. We can see that it fails to match the first "likes" clause but will match the second, if it can find a single constant to instantiate both "X" and "X1". The second "likes" clause is a rule and rules are used to break the solution down to sub-questions: in this instance "human(X) and likes(X,wine).". The question "human(X)" succeeds with the instantiation of "X" by "mary", the second question now becoming "likes(mary,wine).". This matches the first "likes" clause, and so the original question is answered with the instantiation of the variable "X" (and so "X1") to "mary", it so being deduced that "John likes Mary".

From the above description it will be clear that the major problem in understanding PROLOG programs lies not in evaluating the correctness of individual structures or execution steps, but in assessing the long-term implications of a structure or step—it is understanding the teleology of the program that is difficult. Although this is true of many programming languages, it is a very serious problem of PROLOG. This appears to be for three reasons: first, the language has little syntax to guide the user's understanding of flow of control; secondly, the relational structure of clauses encourages an application orientation while programming, which can be in conflict with understanding at the execution level; thirdly, the PROLOG interpreter uses backtracking to seek alternative matches when a match fails, and backtracking has been proven to make programs particularly opaque.

At the top level of description, all clauses starting with the same relation name (be they facts or rules) may be regarded as a procedure. This reduces any account of processing given in an explanation to the order of procedure calls and the entry and exit conditions existing before and after such calls. Given that backtracking is a feature of PROLOG processing, there are four possible parts to a procedure: the initial CALL; a successful return—EXIT; an unsuccessful return—FAIL; a backtrack into a previously completed procedure—REDO (see Fig. 8).

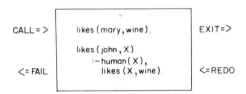

FIG. 8. A top-level view of PROLOG execution.

At the lower level PROLOG is described in terms of the details of processing. This includes a range of operations, the most central of which are the matching of questions (termed "goals") to clauses, the creation of sub-questions ("subgoals"), the instantiation of variables and the process of backtracking.

WORKPAD provides a grammar and vocabulary (listed in a "dictionary" maintained as part of the STRUCTURE GRAPH) for the user to employ for building explanations. The same grammar may be used for explanations at both levels of description. Moreover, levels may be mixed. This is necessary when explaining the execution of a large program, where a detailed exposition will only be required to aid the understanding of areas critical to the solution of a particular problem. As a result of a small amount of experimentation, it has been found that descriptions of PROLOG execution are easiest to write in imperative form. Given that the majority of explanatory statements will concern execution, the explanation grammar requires that both descriptions of structures and events state the relation first, as is standard in predicate calculus. The dictionary may be assessed by the user at any stage of an interaction, and at present contains 47 terms. The grammar itself is very simple and is similar to that described by Sleeman & Hartley (1979). It is given in full below in BNF form.

The grammar states that explanations consist of sets of connected arguments. These arguments are composed of states or event assertions, connected by one of a closed set of relational terms. States and events are also composed from a closed set of terms, all such terms being listed in the system dictionary. An example of an explanation using the grammar is given in Fig. 9. This relates to the simple program presented above concerned with answering the question "Who does John like?". WORKPAD uses indenting to emphasize the structure of arguments.

It is explanations of the form illustrated in Fig. 10 which are presented to the ADVISOR for criticism. The process of generating such critiques will be outlined below.

4.4. THE "ADVISOR"

The ADVISOR's primary role is to guide the user toward a better understanding of the subject domain around the problem currently being addressed via the system. The

```
<explanation>::=<argument>!<explanantion><argument>.

<argument>::=<assertion-list><conclusion-list>.

<assertion-list>::=<assertion>!
                    <assertion-list><assertion>.

<conclusion-list>::=<conclusion-list><conclusion>.

<assertion>::=<state>!<event>.

<conclusion>::=<state>!<event>.

<state>::=<state-relation>'-'<obj-structure-list>.
          /'{'<assert-relation>'}'/.

<event>::=<event-relation>'-'<obj-structure-list>
          /'{'<assert-relation>'}'/!
          <event-relation>/'{'<assert-relation>'}'/.

<obj-structure-list>::=<obj-structure>!
                        <obj-structure-list>','
                        <obj-structure>.

<obj-structure>::=<identifier>':'<PROLOG term>.
```

FIG. 9. WORKPAD explanation grammar.

solving of the problem itself is seen as a side-effect of reaching such understanding. All activity related to the provision of guidance is focussed upon the user explanations in WORKPAD of PROLOG execution and is directed towards identifying both misconceptions and gaps in the user's knowledge. The ADVISOR may be called by the user at any point during the process of explanation building and is called automatically at the end of a program run.

```
            likes1:likes(mary,wine).
            likes2:likes(john,X):-human(X),likes(X,wine).
            human1:human(mary).

QUESTION: ?-likes(john,X1).
RESULT: PROVED likes(john,mary).

**************************************************************

CALL - PROC:likes, GOAL:likes(john,X1).
  MATCH - GOAL:likes(john,X1)., likes1:likes(mary,wine). {RESULT}
  NMATCH
  MATCH - GOAL:likes(john,X1)., likes2:likes(john,X). {RESULT}
  CALL - PROC:human, SUBGOAL:human(X).
    MATCH - SUBGOAL:human(X)., human1:human(mary). {RESULT}
    SUCCEED {AND}
    INSTANT - X:mary
  EXIT - PROC:human
  CALL - PROC:likes, SUBGOAL:likes(mary,wine).
    MATCH - SUBGOAL:likes(mary,wine)., likes1:likes(mary,wine).
                                                          {RESULT}
    SUCCEED
  EXIT - PROC:likes
EXIT - PROC:likes

**************************************************************
```

FIG. 10. A sample explanation.

As stated above, the ADVISOR's objective is to work on user explanations in order to identify understanding difficulties. Once identified, these are:

(a) communicated to the user;
(b) their implications for the current program and explanation are outlined;
(c) the appropriate topics containing corrective information are brought to the user's attention;
(d) the user is invited to make a correction and to re-run the program.

To assist in tracing user difficulties, the ADVISOR may also elicit explanation activity both with reference to the current problem declaration frames and in addition to them.

The principles behind the operation of the ADVISOR are derived from the guidance techniques employed by the most successful human computing advisors observed during the Computing Advisory Service Study reported in section 2. It may be recalled that they used what we described a conservative approach to analysis. Instead of starting their study of user programs and explanations by immediately seeking global patterns of error, they first carefully identified individual points of failure. Only after collecting a number of these did they attempt a unification by mapping them onto organized structures. It may be reasoned that this strategy was adopted because, although the correct identification of some organized pattern of faults gives great problem-solving power, it can be very misleading if such patterns are incorrectly applied. This was especially likely within a programming domain because a given fault can be the product of many different understanding problems.

The MINDPAD ADVISOR adopts a similar strategy. Stored in the STRUCTURE GRAPH are a number of error patterns termed "fault syndromes". These give a description of incorrect patterns of explanation, written in the same explanation grammar as that employed by the user, the whole pattern of faults being itself explained in terms of some particular misunderstanding or lack of knowledge. Also stored in the STRUCTURE GRAPH are descriptions of actions to be taken upon identifying such syndromes.

When critiquing a user explanation, the ADVISOR first proceeds cautiously, testing the validity of each assertion individually with reference to the problem program and its immediate explanatory context. This is done employing the descriptions of PROLOG structures and execution present in the STRUCTURE GRAPH. A number of cycles at this level of analysis may be undertaken. During each pass through the user's arguments, the ADVISOR adds comments to WORKPAD on the current state of the analysis and promote further explanation by the user to test a hypothesis or gain additional evidence concerning the user's understanding problems. At the end of each pass, the ADVISOR attempts to match the current evidence to one of the "fault syndromes" stored in the STRUCTURE GRAPH. If a syndrome is found with an adequate fit, its nature will be explained to the user and remedial action proposed. If a number of different syndromes are identified with a lower level of match, the ADVISOR will now focus on seeking further evidence in order to select one of them with an appropriate level of certainty. If these evaluations fail, the system will then return back to the lower level of criticism.

Factual knowledge within MINDPAD is represented as a set of PROLOG clauses as are both user- and system-generated explanations. It is thus possible to conduct evaluations by transferring clauses between these objects and allowing the ADVISOR to apply the same query types to the resulting knowledge as the user may apply to the topic information.

The analysis of an explanation of a query addressed to an incorrect version of the program given in Fig. 11 will illustrate this process. The initial explanation and program are given below.

```
likes1:likes(john,mary).
likes2:likes(john,X):-human(X),likes(X,wine).
human1:human(mary).
```

```
QUESTION:  ?-likes(wine,mary).
RESULT: PROVED yes              RUN: FAIL no
```

```
***************************************************************

CALL - PROC:likes,  GOAL:likes(wine,mary).
  MATCH - GOAL:likes(wine,mary)., likes2:likes(john,X).
                                           {RESULT}
  INSTANT - X:mary
  CALL - PROC:human,  SUBGOAL:human(mary).
    MATCH - SUBGOAL:human(mary)., human1:human(mary).
  CALL - PROC:likes,  SUBGOAL:likes(mary,wine).
    SUCCEED

***************************************************************
```

FIG. 11. An example of an incorrect and incomplete explanation.

Given the above explanation, the ADVISOR makes its first scan through the text and notes that it is incomplete at the top procedural level. There are three procedure CALLs but no EXITs or FAILs. It thus begins the analysis by parsing for possible procedure exits and marks these in the explanation. To achieve such parsing, the ADVISOR must make reference to constraints operating at the lower level of PRO-LOG description. With the present example, this would provide the additional information that an incorrect result of the MATCH operation had been declared on the first call. It is also noted that the user has failed to indicate that PROLOG would first attempt a match to likes1, which would fail. However, this would not necessarily indicate that the user did not understand the matching process: he could have simply decided to abbreviate the explanation. As a result of this analysis the following annotations are written into the explanation text and denoted by angle-brackets (Fig. 12).

Having made a preliminary analysis, the ADVISOR relates the current evidence to fault syndrome descriptions in the STRUCTURE GRAPH, and finds that none of them match adequately. It thus continues by attempting to identify some important source of user error from the commented explanation which it can use to plan further action. In the present instance, the evidence points to a lack of understanding concerning the topic MATCH. The user is therefore directed to read this and to amend the explanation from what is learned (see Fig. 13).

The ADVISOR now works on this amended explanation. It is again incorrect, although, there is additional evidence concerning the understanding difficulty. First, the user has now included correctly the initial failed match with likes1. However, given a complete understanding of the topic, there is a contradiction with the result of the match with likes2. Secondly, the final match is not declared as being between a subgoal

```
****************************************************************
<<<Missing MATCH to likes1; possible abbreviation>>>
CALL - PROC:likes, GOAL:likes(wine,mary).
  MATCH - GOAL:likes(wine,mary)., likes2:likes(john,X).
                                            {RESULT}
    INSTANT - X:mary
<<<MATCH would not SUCCEED>>>
<<<Likely FAIL; suspect user considers EXIT>>>
  CALL - PROC:human, SUBGOAL:human(mary).
    MATCH - SUBGOAL:human(mary)., human1:human(mary).
<<<Missing indication of whether MATCH SUCCEEDS; suspect
   user considers SUCCEEDs>>>
<<<On previous assumption, assume user considers EXIT>>>
  CALL - PROC:likes, SUBGOAL:likes(mary,wine).
    SUCCEED
<<<No indication of the HEAD MATCHed; SUBGOAL does not MATCH
   any FACT or RULE, therefore the CALL would FAIL>>>

****************************************************************
```

FIG. 12. ADVISOR's preliminary analysis.

```
****************************************************************
CALL - PROC:likes, GOAL:likes(wine,mary).
  MATCH - GOAL:likes(wine,mary)., likes1:likes(john,mary).
  NMATCH
  MATCH - GOAL:likes(wine,mary)., likes2:likes(john,X).
  INSTANT - X:mary
  CALL - PROC:human, SUBGOAL:human(mary).
    MATCH - GOAL:human(mary)., human1:human(mary).
    SUCCEED
  EXIT
  CALL - PROC:likes, SUBGOAL:likes(mary,wine).
    MATCH - SUBGOAL:likes(mary,wine)., GOAL:likes(wine,mary).
    SUCCEED
  EXIT - PROC:likes
EXIT - PROC:likes

****************************************************************
```

FIG. 13. User's amended explanation.

and a clause head, but between a subgoal and goal. Thirdly, it is declared as being successful. Even if one of the elements was a clause head, the match would have failed because the constants in the arguments of the two predicate structures are reversed.

This pattern of evidence is sufficiently close to one of the fault syndromes to warrant the direct testing of the user's understanding. The test concerns questioning why the initial match with likes1 failed, while that with likes2 succeeded. The reply is that the variable "X" needed to be instantiated with the value "mary". Asked why this was required, the user points out that it was needed so that the "X" of the final predicate structure could take the value "mary". In this case, the replies, when taken with the attempt to match a goal to a subgoal indicated that the user did not understand the topics HEAD, GOAL and SUBGOAL, and therefore did not properly understand this aspect of the MATCH. As a result of this, he had made the common error of attempting to work a program backwards from an assumed match, in this case with the final structure of likes2.

After a fault syndrome has been selected, the ADVISOR instructs the user to correct the difficulty identified. Following this, the user is asked to assess the validity of the conclusion, and to make appropriate changes to the explanation. If these are incorrect, the analysis process continues with the focus on detail again. In the present instance, the user gave the explanation presented in Fig. 14 which records in detail the correct execution record of the query.

```
***************************************************************

CALL - PROC:likes, GOAL:likes(wine,mary).
  MATCH - GOAL:likes(wine, mary)., likes1:likes(john,mary).
  NMATCH
  MATCH - GOAL:likes(wine,mary)., likes2:likes(john,X).
  NMATCH
FAIL - PROC:likes

***************************************************************
```

FIG. 14. A correct record of program execution.

5. Steps to implementation

We have been very sketchy in indicating implementation details such as the precise nature of critiquing heuristics to be used by the advisor. These have been omitted intentionally from the present discussion because our interest is in introducing computer-based guidance systems as a general class of knowledge-based program. Individual implementations may differ in the actual facilities provided, the level of program control, intelligence and the range of application.

With reference to the PROLOG guidance system, implementation decisions are based as far as possible on experiment. As mentioned in section 4, Pask has already implemented a number of tutorial systems within the bounds of Conversation Theory. The approach appears therefore to be viable, at least in its passive form. However, there are sufficient differences between a tutorial and a guidance context to proceed cautiously. Instead of attempting to build a prototype system first off, a range of experiments will be run to help determine the design of such features as the explanation cycle and the explanation grammar. With reference to the latter, great attention will have to be given to user friendliness because users may be unwilling to learn what amounts to a formal language in order to come to understand a second formal language.

Our own approach involves the simulation of the computer consultation with the experimenter in the role of the system, progressively incorporating the computer-based facilities as they are developed. Such facilities include WORKPAD for the building of explanations, and the STRUCTURE GRAPH as a database of information on PROLOG. It is also possible to use a prototype ADVISOR under the control of the experimenter to offer suggestions when the user declares that he is facing difficulty. Finally, it is expected that subject-specific elements such as the choice of appropriate learning and analysis heuristics to be employed by the ADVISOR, and the nature of "fault syndromes" will emerge during experimentation. The guidance system is therefore being incrementally validated during construction, rather than waiting until it is a completed product when it is difficult to make even moderate changes in design.

The MINDPAD design section of this paper was written while the first author was visiting the University of Pittsburgh during the Winter Semester, 1983. Thanks are due to Jim Reggia for a detailed critique of early drafts and Jim Greeno and Harry Pople for a number of formative discussions over this period. Their encouragement was much valued, as were their comments on the viability of the ADVISOR design. Thanks are also due to the Department of Computer Science, University of Pittsburgh for their hospitality, and to the Department of Computer Science, University of Strathclyde for giving the first author leave of absence to develop the MINDPAD design. The Advisory Service Study reported in the first part of the paper was supported by Social Science Research Council grant number HR 4421.

References

ALTY, J. L. & COOMBS, M. J. (1980). Face-to-face guidance of university computer users—I: a study of advisory services. *International Journal of Man–Machine Studies*, **12**, 390–406.

ALTY, J. L. & COOMBS, M. J. (1981). Communicating with university computer users: a case study. In COOMBS, M. J. & ALTY, J. L., Eds, *Computing Skills and the User Interface*. London: Academic Press.

BARTLETT, F. C. (1932). *Remembering: A Study in Social and Experimental Psychology*. Cambridge: Cambridge University Press.

BRUNER, J. S. (1957). Going beyond the information given. In GRUBER, H., HAMMOND, K. R. & SESSER, R., Eds, *Contemporary Approaches to Cognition*. Cambridge, Massachusetts: Harvard University Press.

BRYD, L. (1980). Understanding the control flow of PROLOG programs. *D.A.I. Research Paper No. 151*, University of Edinburgh.

CLANCEY, W. J. (1979). Tutoring rules for guiding a case method dialogue. *International Journal of Man–Machine Studies*, **11**, 25–49.

CLANCEY, W. J. & LETSINGER, R. (1981). NEOMYCIN: Reconfiguring a rule-based expert system for application to teaching. *IJCAI-7*, 829–836.

CLOCKSIN, W. F. & MELLISH, C. S. (1981). *Programming in PROLOG*. Berlin: Springer-Verlag.

COOMBS, M. J. & ALTY, J. L. (1980). Face-to-face guidance of university computer users—II: characterising advisory interactions. *International Journal of Man–Machine Studies*, **12**, 407–429.

COOMBS, M. J. & ALTY, J. L. (1982). An application of Sinclair's discourse analysis system to the study of computer guidance interactions. *Journal of Human–Computer Interaction* (submitted).

COOMBS, M. J. & HUGHES, S. (1982). Extending the range of expert systems: principles for the design of a consultant. Paper presented at *Expert Systems 82*, Brunel University.

DAVIS, R. (1983). Reasoning from first principles in electronic troubleshooting. *International Journal of Man–Machine Studies*, **19**, 403–423.

DAVIS, R. & LENAT, D. (1982). *Knowledge-based Systems in Artificial Intelligence*. New York: Academic Press.

FEIGENBAUM, E. A., BUCHANAN, B. G. & LEDERBERG, D. J. (1971). On generality and problem-solving: a case study using the DENDRAL program. In MELTZER, B. & MICHIE, D., Eds, *Machine Intelligence 6*. Edinburgh: Edinburgh University Press.

GREGORY, R. L. (1970). On how so little information controls so much behaviour. *Ergonomics*, **13**, 25–35.

KELLY, G. A. (1955). *Psychology of Personal Constructs I*. New York: Norton.

KULIKOWSKI, C. A. & WEISS, S. M. (1982). Representation of expert knowledge for consultation: the CASNET and EXPERT projects. In SZOLOVITS, P., Ed., *Artificial Intelligence in Medicine*. Boulder, Colorado: Westview Press.

LEHNERT, W. G. (1978). *The Process of Question Answering*. Hillsdale, New Jersey: Lawrence Erlbaum.

LENAT, D. B. & HARRIS, G. (1978). Designing a rule system that searches for scientific discoveries. In WATERMAN, D. A. & HAYES-ROTH, F., Eds, *Pattern Directed Inference Systems*. New York: Academic Press.

NEWELL, A. & SIMON, H. A. (1972). *Human Problem Solving.* Englewood Cliffs, New Jersey: Prentice–Hall.

PARKER-RHODES, F. (1978). *Inferential Semantics.* Sussex: Harvester Press.

PASK, G. (1975). *Conversation, Cognition and Learning.* Amsterdam: Elsevier.

PASK, G. (1976). *Conversation Theory: Applications in Education and Epistemology.* Amsterdam: Elsevier.

PIAGET, G. (1972). *Psychology and Epistemology: Towards a Theory of Knowledge.* London: Penguin University Books.

SCHANK, R. C. (1979). Reminding and memory organization: an introduction to MOP's. *Research Report 170.* Department of Computer Science, Yale University.

SCHANK, R. C. & ABELSON, R. (1977). *Scripts, Plans, Goals and Understanding: An Inquiry into Human Knowledge Structures.* Hillsdale, New Jersey: Lawrence Erlbaum.

SHORTLIFFE, E. H. (1976). *Computer-based Medical Consultations: MYCIN.* New York: Elsevier/North–Holland.

SLEEMAN, D. H. & HARTLEY, R. J. (1979). ACE: a system which analyses complex explanations. *International Journal of Man–Machine Studies,* **11**, 125–144.

SUSSMAN, G. J. (1975). *A Computer Model of Skill Acquisition.* New York: American Elsevier.

VAN DIJK, T. A. (1980). *Macrostructures: An Interdisciplinary Study of Global Structures in Discourse, Interaction, and Cognition.* Hillsdale, New Jersey: Lawrence Erlbaum.

Knowledge reorganization and reasoning style

CHRISTOPHER K. RIESBECK

Department of Computer Science, Yale University, New Haven, Connecticut, U.S.A.

To study the learning of expertise, two closely related stages of expertise in economics reasoning are analyzed and modelled, and a mechanism for going from the first to the second is proposed. Both stages share the same basic concepts and generate plausible economic scenarios, but reasoning in the first stage oversimplifies by focussing on how the goals of a few actors are affected. Reasoning in the second stage produces better arguments by taking into account how all the relevant parts of the economy might be affected. The first stage is modelled by highly interconnected goal forests and very selective, story understanding search heuristics. The second stage is modelled with more explicit links between economic quantities and a more appropriate set of search heuristics. The learning mechanism is a failure-driven process that not only records better arguments as they are seen, but also records the failure of existing inference rules to find these arguments on their own. The collected failures are used to determine which search heuristics work best in which situations.

Introduction

Expert systems, such as PROSPECTOR (Duda, Hart, Nilsson & Sutherland, 1978) and MOLGEN (Stefik, 1981), are clearly very useful in dealing with well-defined knowledge-intensive real-world tasks, and the technology for building them is improving constantly. But it is important to step back and realize that in some ways expert systems are crucially different from human experts. Perhaps the most important difference is that, except in the area of classification, most expert systems were never novices. That is, they didn't start with one kind of knowledge base and learn a new one.

Potentially an expert system that had been trained, that is, that had learned its expert knowledge starting with a non-expert knowledge base, would have several advantages.

It could use its non-expert knowledge to explain its decisions to non-experts.
It could use its non-expert knowledge to handle unexpected situations, because such knowledge, though less effective, tends to be more general.
It could cope with changing domains, by continuing to update its knowledge base.

Unfortunately, developing a novice model that can learn involves at least three very hard problems:

how to learn rules and facts;
how to learn new concepts; and
how to reorganize knowledge.

Research continues on all three fronts. People are working on how new rules and facts are understood and integrated into the structures that are already in memory (Schank, 1982; Lebowitz, 1980; Kolodner, 1980; Norman & Bobrow, 1977). Others are working on how new concepts can be formed by induction (Michalski, 1981;

159

DEVELOPMENTS IN EXPERT SYSTEMS
ISBN 0-12-187580-6

Mitchell, 1978; Winston, 1979) or from analogies (Gentner, 1982; Carbonell, 1979; Burstein, 1981).

This article is about the current research being done by the Yale Economics Learning Project into the third aspect of learning to be an expert, the reorganization of knowledge. The following sections discuss the nature of the economic domain, present some examples of expert and novice reasoning, analyze some of the differences, propose an AI model for each kind, and outline a mechanism for getting from the novice model to the expert. Although the Learning Project has been going for several years, all of this is still very much work in progress.

Advantages of the economics domain

We selected the domain of economics to study the learning of expertise quite deliberately. It may seem strange to study learning expertise in a domain notorious for how much its experts disagree. But, in fact, this constant disagreement is one of the things we were looking for, because we are interested in *learning from failure*. One of the nice things about economics is that any reasonable argument one might believe has several equally reasonable counterarguments to convince you otherwise.

It is also important that economics is an everyday topic among non-experts. We couldn't study the reorganization of economic knowledge unless novices had some economic knowledge to start with. Furthermore, economics as it is discussed in the daily paper is quite accessible to the average person. Many people can understand most of the arguments that politicians, economists, and editorial writers present in the pages of the *Wall Street Journal, Time* magazine, and so on.† However, the average person is not as good at coming up with these arguments. Part of being an expert in this domain is an increased ability to argue and counter-argue.

One final positive feature of economic reasoning is that it appears to be strongly rule-based. Hence, the development of economic reasoning is relevant to the learning of rule-based expertise in general.

Some examples of economic reasoning

In order to study expert versus novice reasoning, we need some examples of each. Expert reasoning (at least, the kind of expert we are interested in) is easily found, in the pages of the *Wall Street Journal,* the *New York Times, Newsweek,* and so on. Fortunately for us, examples of novice reasoning have also become available, thanks to William Salter (1983). The protocols he has gathered have been invaluable.

Consider the following examples of economic reasoning.

> With the resulting structure of taxes and expenditures, the President is not going to be balancing the Federal budget in 1984 or any other year. With high growth choked off by high interest rates, budget deficits are going to be bigger, not smaller. The result: more demands for credit and higher interest rates.
>
> Lester Thurow, *Newsweek,* 21 September 1981, p. 38

> Q: If interest rates went up, what would happen to the federal budget deficit?

† Economics as it is taught in college is a different story, where there is a rapid introduction of new concepts and rules.

A: Increase; business can't afford capital improvements; individuals do not have the credit they need for new homes, appliances, cars. This means an increase in inflation but it also means that the amount of money businesses would make benefitting from this investment would not be there. Therefore, taxes coming into the government are going to be lower; therefore, there is going to be a bigger amount of shortfall.

> Subject 1, protocol taken by William Salter

Q: If the federal budget deficit went down, what would happen to the rate of inflation?
A: I think inflation would go down, because people would have less money taken out of their weekly income, they hopefully would do some more saving, have more money to live on, not be pushing for higher wages, and that wouldn't be pushing inflation up.

> Subject 2, protocol taken by William Salter

Q: If government spending went down, what would happen to the rate of inflation?
A: Well, it should go down. (HOW COME?) Hmm... that's a good one... how come....I'm just going to leave that one blank, OK?

> Subject 21, protocol taken by William Salter

I hope that reader sees a steady decrease in expertise in these examples of reasoning by four different people. The first and last are the most extreme. Thurow, a well-known economist, produces a coherent argument, while Subject 21 has a belief, but is unable to justify it. The difference between Thurow and the other three examples is exaggerated by the fact that the Thurow quote was written (and probably rewritten) while the others were verbal. Even so, the difference is nowhere near as great as the difference that would be seen if we compared a theoretical physicist reasoning about the atom with a non-physicist. Most likely, even Subject 21 would able to follow Thurow's argument, even though Subject 21 would not be able to come up with it.

The second and third protocols, which we will focus on from this point forward, are much closer together, but I want to claim that these two protocols use different reasoning strategies. In fact, I want to make a number of speculative claims.

> The second protocol is a search through a graph of directed signed connections between economic quantities.
> The third protocol is a search through a hierarchy of the goals and subgoals of economic actors.
> The second protocol is closer to expert reasoning, while the third is closer to novice reasoning.
> Expert reasoning is not radically different from novice reasoning, but the result of gradual distributed changes to novice reasoning.
> The mechanism of change is the addition of new links in response to self-acknowledged failures in reasoning.

The next few sections expand upon these claims.

Link-based search

When you look at economic reasoning, one kind of argument stands out: one economic quantity affects another, because of a chain of effects on intermediate quantities. For example, high interest rates affect inflation by affecting business investment which affects productivity which affects prices which affects inflation.†

---

† Other ways to reason about the economy are analogies ("Money is the fuel for economic activity") and examples ("Large Federal deficits in the U.S.A. accompanied inflation in both World Wars").

THE GRAPH MODEL

An obvious model for this kind of reasoning is graph search, where the graph has nodes standing for economic quantities and links standing for the effects quantities have on each other. The elements of this kind of model for economic reasoning are:

the economic quantities (e.g. inflation, unemployment, the federal deficit);
the net effects between them (e.g. interest rates negatively affect investment); and
the search algorithm for finding connections between any two quantities.

The links are signed and directed to represent how one quantity affects another. For example, to represent the belief that investment increases productivity, we would add a positive link from the node representing investment levels to the node representing productivity. To represent the belief that higher interest rates cause investments to decrease, we would add a negative link from interest rates to investment.

A signed link between two quantities actually represents, crudely, two inferential relationships between those quantities.

Given an increase or decrease in the first quantity, infer an appropriate change in the second quantity.
Given a high or low level for the first quantity, infer an appropriate level for the second quantity.

Thus if interest rates are high, a negative link from interest rates to investments would imply that investment levels are low. If the interest rates are falling, the same negative link would imply that investments would be increasing.

Signed links can also occur between a quantity and another link, representing the effect one quantity has on how much a second quantity affects a third. For example, later we will need a positive link from consumer spending to the link from investment to profits, to represent the belief that the level of consumer spending positively affects returns on investments.

A chain of effects is represented as a sequence of links. If node A is linked to node B, and node B is linked to node C, then node A is linked indirectly to node C. The sign of the indirect link from A to C is the product of the signs of the links from A to B and B to C, that is, the sign is positive if and only if A to B and B to C have the same sign, and the sign is negative otherwise. In Fig. 1, the node A is positively connected to D through B and negatively connected to D through C.

There are many more things to be added to such a representation scheme before it can be considered adequate enough for economic reasoning. Links have many properties. If we have a link from A to B, we need to give its size (i.e. how big an effect A has on B), its certainty (i.e. how likely A is to affect B), its delay (i.e. how long it takes before a change in A affects B), and so on.

FIG. 1. Example graph of signed directed links.

Furthermore, links can be added or subtracted (e.g. a budget is the difference of income and outgo), integrated or differentiated (e.g. inflation is the rate of change in prices), or measured (e.g. interest rates measures the size of the negative "cost" link from borrowing to borrower's money supply). The reader is referred to the rapidly growing literature on qualitative reasoning about quantities (deKleer & Brown, 1982; Kuipers, 1982; Forbus, 1981).

GRAPH SEARCH

A graph alone does not a model of reasoning make. We also have to specify how the graph is searched to answer the question "How does quantity X affect quantity Y?" There is a large literature on graph search [see Nilsson (1980) for a good introduction], and there are many options.

First, consider a bare-bones template of a forward-looking graph searcher whose job is to find a connection between some node X and another node Y. Such a searcher keeps track of the nodes in the graph it has reached, plus some information about how it got to those nodes. The searcher picks a node with links leading from it that haven't yet been followed, and follows some of those links to the nodes at the other end. If any of these nodes is Y, the searcher has found a path from X to Y. Otherwise, the searcher picks another node and tries again. This template can become a large number of possible graph searching algorithms, depending on how much information is saved when a node is reached, and what rules are used to pick nodes from which to follow links.

Basically, there are two kinds of rules for picking nodes. One kind picks the node that has the "best" path so far. "Best" might mean quickest, strongest, most certain, shortest, or whatever. These rules try to find the best path from X to Y. They require that the path to a node be saved when a node is reached, and that when the searcher finds a new path to a node it already has a path to, it keeps the shorter one for future use.

The other kind of rule for picking nodes takes the one that seems closest to the goal node. For example, a node might be judged close to the goal node if it is in the same knowledge structure as the goal node or shares a number of features with the goal node. These rules try to find a path from X to Y as quickly as possible. They do not require information about paths to the node that have been found so far, and need not generate the best path between X and Y.

GRAPH SEARCH EXAMPLE

We can now take the protocol from Subject 1 on p. 47, repeated below, and start filling in some of these options in the graph search template. Subject 1 was one of the more expert of Salter's novices, having earned a Master's degree in economics a number of years earlier.

> Q: If interest rates went up, what would happen to the federal budget deficit?
> A: Increase; business can't afford capital improvements; individuals do not have the credit they need for new homes, appliances, cars. This means an increase in inflation, but it also means that the amount of money businesses would make benefitting from this investment would not be there. Therefore, taxes coming into the government are going to be lower; therefore, there is going to be a bigger amount of shortfall.
>
> Subject 1, protocol taken by William Salter

I'll assume that there are three chains here:†

interest rates to investment;
interest rates to spending to inflation; and
interest rates to spending to investment returns to tax revenues to the deficit.

A subgraph for Subject 1's protocol appears in Fig. 2. Note that there is a link from a quantity (spending) to a link (from investments to returns).

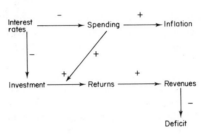

FIG. 2. Subgraph for second protocol.

It appears that the subject's search algorithm is not trying to find the shortest path. Why? Because the subject picks the chain from decreased spending to decreased investment returns, even though the subject had previously concluded that investment itself would be reduced. In other words, the subject could have said, given the subject's own protocol, that higher interest rates reduce investment, which reduces returns, which reduces tax revenues and increases the deficit. It seems reasonable to assume that Subject 1 was trying to find a path as quickly as possible, not as short as possible.

To determine Subject 1's heuristics for picking links to follow, we must go through the subject's entire protocol, which has answers to 24 questions from one session and 24 questions (12 old, 12 new) from a session a month later, and see if a consistent set of heuristic rules can be devised. This is part of the work in progress now. This task is unlikely to fail since no *a priori* limits on heuristics have been set. If, however, there is no convergence on the set of heuristics after several expert protocols have been studied, then the graph model is probably inappropriate for this stage of reasoning.

Goal-based reasoning

Although graph searching seems reasonable for modelling Subject 1, expert economic reasoning, I now want to claim that it is a bad model for novice economic reasoning. Here again is the protocol from Subject 2 on p. 47.

> Q: If the federal budget deficit went down, what would happen to the rate of inflation?
> A: I think inflation would go down, because people would have less money taken out of their weekly income, they hopefully would do some more saving, have more money to live on, not be pushing for higher wages, and that wouldn't be pushing inflation up.
>
> Subject 2, protocol taken by William Salter

† The protocol is ambiguous, in that the subject might have meant either that inflation hurts investment returns or that decreased consumer spending hurts inflation. By assuming three chains we don't have to worry about whether Subject 1 really believed that less credit spending is inflationary.

If this protocol were modelled with graph search, it would require nodes such as "push for higher wages" and "have money for living", plus links such as the negative link from "have money" to "push for higher wages". These nodes are not quantitative measurements of the economy. They are goals and plans that people have. The negative link from "have money" to "push for higher wages" is an instance of a much more general connection, namely that someone doesn't need to ask for something if they already have it.

A better model of the reasoning above, and of novice reasoning about the economy in general, is to view it as goal-and-plan problem-solving. The reasoning still involves finding paths between nodes, but now the nodes represent the goals of economic actors and the links are inferred from the relationships between goals.

The elements of goal-based economic reasoning are:

the actors (government, business, consumers, labor, banks);
the goals of each actor (balanced budget, high profits, consumables, etc.);
the subgoals used to achieve the goals (raise taxes, increase sales, borrow money, etc.);
the controllable goals of each actor (consumers control rate of sales, businesses control prices, etc.); and
the algorithm for finding connections between economic quantities, by finding connections between affected goals.

Our model of economic reasoning oversimplifies certain aspects of goal-based reasoning. In particular, goals, plans, and actions are lumped into one category called a "goal activity" or simply "goal". For example, a consumer's goal of INCREASE MONEY is both the goal of having money and the activity of getting money, and it has a plan which is the goal activity of DECREASE SPENDING.

Complexities arise because the same goals often appear in more than one goal tree. Normally therefore we have to deal with goal *forests*. Figure 3 has an abstract example. The goal of increasing A is a subgoal of two goals: increasing B and increasing D. B and D might be goals for the same actor or for separate actors.

FIG. 3. Abstract goal forest example.

GOAL-BASED LINKING

When a goal-based reasoner is asked "How does the economic quantity X affect the economic quantity Y?" the searcher has to find a path from X to Y using two kinds of links:

links between goals and economic quantities and
links between goals.

The first kind of link can be direct or through the definition of the quantity. For example, there is a direct link from inflation to the government's goal DECREASE INFLATION. But, since inflation measures increases in how much money changes hands when someone buys something, inflation is also linked, by definition, to the goals of consumers and business.

The second kind of link is derived from the connections between subgoals and goals. There are two simple rules for making these links.

Rule of ASCENT. If SG is a subgoal of G, then SG is linked positively to G.
Rule of DESCENT. If SG is a subgoal of G, then G is linked negatively to SG.

The ASCENT rule is obvious. A subgoal is a subgoal in part because it is positively connected to its main goal. In the DESCENT rule, the connection from main goal to subgoal is negative, because if the main goal is already being satisfied, then the subgoal is not needed, while if the main goal is being satisfied less, then the subgoal is needed more.

The DESCENT rule has two problems. First, even though a main goal is satisfied, the actor might want even more. Second, since there are usually several subgoals to a main goal, it is not certain which subgoals will be affected by a change in the main goal.

For these reasons, I believed for some time that the DESCENT rule would be used very sparingly. However, the protocol from Subject 2 on p. 47 and below, uses the DESCENT rule no less than three times! The partial goal forest involved is in Fig. 4.

> Q: If the federal budget deficit went down, what would happen to the rate of inflation?
> A: I think inflation would go down, because people would have less money taken out of their weekly income, they hopefully would do some more saving, have more money to live on, not be pushing for higher wages, and that wouldn't be pushing inflation up.
> Subject 2, protocol taken by William Salter

Figure 4 shows trees for three actors, government, labor, and business, and the DESCENT rule is used once in each. First, the government, with a reduced deficit, can lower taxes. Second, the wage earners, with reduced taxes, can reduce salary demands. Third, the businesses, with non-increasing wages, won't have to increase prices.

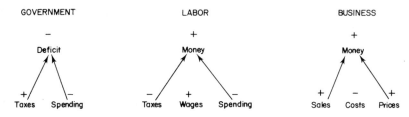

FIG. 4. Goal graph for Subject 2's protocol.

There are at least two possible reasons why the DESCENT rule should appear so frequently here.

Some of the links have been previously stored and no longer derived by explicitly applying the DESCENT rule. In particular, the link from wages to prices and the

link from deficit to taxes, which are commonly talked about in the press, are good candidates for stored links.
Some of the links are involved in trade-off situations.

Both explanations are plausible. I will focus on the second, not because it is necessarily more likely in this case, but because it is theoretically more interesting.

Trade-off situations result from goals in conflict or competition (Carbonell, 1979; Wilensky, 1978). A trade-off exists when some activity positively affects one goal but negatively affects another. As a result, the activity is done at a rate which balances the good with the bad. For example, consumer spending is a balance between having objects and losing money. Price setting is a balance between consumer demand and business profits.

Trade-off situations are important in economic reasoning because of two basic assumptions:

at any given time, all trade-offs are at equilibrium and
if disturbed, trade-off situations return toward equilibrium.

For example, business investment is a trade-off. It costs money but increases productivity. It is assumed that businesses are investing at some level that balances productivity with cash-on-hand. If a tax cut for businesses gives them more money, then a new balance is reached, shifting some of the money from the tax cut over to investment. Conversely, if taxes increase, decreasing cash-on-hand, then businesses will decrease investment.

In Subject 2's protocol, at least two uses of the DESCENT rule involve trade-off situations. First, taxes are a trade-off for the government, which balances the need for them against their unpopularity with voters. Second, prices are a trade-off for businesses, which make them high enough to make a profit but low enough to keep sales going. Only the use of the DESCENT rule to conclude that workers with more money will not demand wage increases does not involve a direct trade-off situation. This may be why Subject 2 used the word "hopefully" when there were no such qualifications on the other two uses of the DESCENT rule.

GOAL-BASED SEARCH HEURISTICS

The rules given for constructing links between goals imply that the novice reasoning model faces a very serious combinatorial explosion in search time with even a few goal-trees

because every goal is connected to many others, either
in conflict or competition with other goals, or
because the same person or group can be viewed as more than one economic actor; and
because the ASCENT and DESCENT rules allow the searcher to travel up and down the goal trees, connecting *any* pair of goals, as long as the goal trees involved have at least one common goal.

In other words, the combinatorial problem for goal-based linking is almost as bad as search in a completely connected graph. Hence, the novice searching algorithm has to have very highly selective heuristics.

The following classes of search heuristics are currently part of the model.

ASCENT. Prefer ASCENT links to DESCENT. Currently, only one use of the DESCENT rule not justified by another heuristic is allowed.
TRADE-OFF. If the changed goal is part of a trade-off situation, follow the trade-off links.
PARITY. Prefer good effects from good changes, and bad effects from bad changes.
HIERARCHY. Prefer links to higher-ranked goals.
HILL-CLIMB. Prefer links to goals that share the most features with the target goal.

In order for the PARITY heuristic to work, there must be some independent determination of the goodness of economic quantities, since it would be circular to say that a quantity is good if it has good effects. Fortunately, it seems reasonable to assume that many quantities, such inflation, taxes, unemployment, and the deficit, have been clearly labelled by politicians and the media.

The PARITY heuristic seems to be a very strong force in Salter's protocols. Sometimes there appear to be fairly obvious mistakes in the signs of links, and when they occur, these mistakes preserve rather than violate parity. Also, there are times when subjects can give an answer, but not a reason, as in:

> Q: What would happen to inflation if the federal budget deficit went down?
> A: Well, inflation should go down also. (WHY?) For the simple reason that . . . I'm trying to think . . . . Can we come back to that one? I just can't . . . figure it out.
> Subject 17, protocol taken by William Salter

In most of these cases, the answers given preserve parity.

The HIERARCHY heuristics are based on Carbonell's (1979) model of ideologies as hierarchies that specify, for a reasoner, which goals are most important to that reasoner, and which goals that reasoner thinks are most important for other actors. A right-wing conservative, for example, might rank national security as very important, and believe that Communists primarily want to conquer the world. A left-wing liberal might rank unemployment as very important, and believe that Communists just want to be left alone.

Ideologies clearly affect novice economic reasoning. A consumer-oriented view holds that the ability of consumers to buy things is most important, and that businesses are totally self-interested. For them, the important effect of high interest rates is their detrimental affect on consumption, and they reason that reducing business taxes is just a ploy to make businessmen rich.

Subject 2's chain of reasoning in the protocol went from the deficit to wages to inflation. Subject 2 consistently went to workers and consumers in reasoning. Interest rates affected consumption, business taxes affected salaries, and productivity and government spending affected jobs. Hence, a model of Subject 2's ideology would be to rank consumer and labor goals over business and government ones.

These ideological rankings guide the goal forest search algorithm in two ways. First, when using the ASCENT rule on a subgoal that appears in more than one goal tree, the ranking says which actor's goals are most important and should be followed. Second, when using the DESCENT rule on a goal which has more than one subgoal, the goal hierarchy says which subgoal is most likely to be increased or decreased.

Teaching by argument

Part of our research is the development of these two simple models of economic reasoning, goal-based linking for novices and link-based searching for experts. The other part is a theory of learning which can start with goal-based linking and develop into link-based searching.

The methodology is simple enough. A goal-based linker is given two kinds of input: arguments and questions. The arguments are claims, such as "High interest rates are inflationary", and a sequence of supports, such as "They slow investment, decreasing productivity, thereby raising prices". Importantly, some arguments contradict others, such as "High interest rates are deflationary, because they restrain spending, which lowers prices". The questions ask about link between quantities. The program's answers should change as the program learns from its mistakes.

There are some fairly objective constraints on what would make a reasonable learning program.

There must be some arguments that the program will accept.
There must be some arguments that will change the program's mind. That is, given two contradictory arguments, A and B, after A, it believes A, but after B it believes B.
There must be some arguments that don't change the program's mind. That is, given two contradictory arguments, A and B, after A, it believes A, and after B it still believes A.
There must be no arguments that the program will oscillate between forever. That is, it must not happen that, given two contradictory arguments, A and B, after A, it believes A, after B it believes B, after A, it believes A again, and so on.

There are also some more subjective reasonability constraints on the program.

The program must be able to give reasonable arguments.
The program's beliefs must be reasonably consistent with each other.
The program must learn faster than one argument at a time.
The program's arguments should depend less and less on goals and the PARITY and HIERARCHY heuristics.

The first and second conditions are initially met by the novice reasoner. Hence, the requirements are that learning should not interfere with reasoning. The third condition says that learning is more than memorizing facts or individual arguments. The fourth condition says that the learning should cause the reasoning to change in the direction suggested by the protocols and published articles.

The next section outlines the learning mechanism, and the section after that argues why this mechanism may satisfy the conditions just given.

Failures and exceptions

Our learning mechanism is triggered when the economic reasoning model reads an argument better than the one it comes up with itself. This causes two things to be added to memory:

an explicit link representing the best path to follow, and why; and
an explicit structure representing the failure of the heuristic that took the wrong path.

The basic structure for adding new links and recording failures is the *Exception Memory Organization Packet* (Riesbeck, 1981), which is one example of the Memory Organization Packet (or MOP) record-keeping and generalization structure (Schank, 1982; Kolodner, 1980; Lebowitz, 1980). MOPs are modular frame structures that organize, index, and cross-reference episodic knowledge of events. MOPs are stored in a discrimination net with two basic kinds of links: abstraction (i.e. one MOP is a more general case of another) and packaging (i.e. one MOP has several other MOPs as subparts).

One special kind of episode is an internal processing failure. Exception MOPs package these failures into the following subparts:

the FAILURE, which is an error message constructed by some process;
the EXPLANATION, which is a rule that is blamed for the failure; and
the CORRECTION, which is a rule that should have been used instead of the EXPLANATION.

In the simple economics reasoner, we currently have only one kind of failure: BETTER-CHAIN-FAILURE. A BETTER-CHAIN-FAILURE means that the input argument is a chain that is better (stronger, shorter, more certain) than the reasoner's self-generated path. For our purposes, it doesn't matter whether the two paths agree in sign, but this would be very important for a reasoner with a vested interest in certain effects going in certain directions.

Here is an abstract example of how Exception MOPs are formed. Assume the reasoner has the goal forest in Fig. 3 and the rules in Fig. 5. The rule R-1 is used to connect A with another goal, and the heuristic H-1 ranks the B goal as important. H-1 supports the default action of R-1 to go from A to B rather than from A to C.

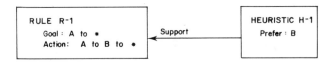

FIG. 5. Example rules before failure processing.

Suppose now that the input is "A positively affects D, because A positively affects C which positively affects D". Because of the heuristic H-1, the reasoner would generate a different path to D, namely "A negatively affects D, because A positively affects B which negatively affects D". The link from B to D requires the DESCENT rule.

The input chain "A to C to D" is better than the reasoner's because it doesn't use the DESCENT rule. Therefore, the reasoner has a BETTER-CHAIN-FAILURE and creates an Exception MOP, E-1, with the EXPLANATION set to H-1, and the CORRECTION set to a new rule R-2, which is supported by this episode (see Fig. 6).

Becoming less novice

The first comment to make about Fig. 6 is that the new rule R-2 contains more than just a link from goal A to C. It has the whole path from A to D. Furthermore, the

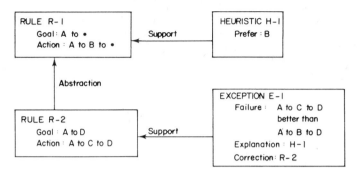

FIG. 6. Example rules after failure processing.

SUPPORT for R-2 is connected to E-1, which contains the old path through B. Hence R-2 is a direct link from A to D, underneath which can be found the two paths on which it is based.

Traces of the search for a connection from A to D before and after the failure would be quite compatible with claims made earlier about the difference between novice and expert protocols. Before the failure, the trace of rule applications would be "An increase in goal A leads to an increase in goal B, and that probably would cause a decrease in goal D". The "probably" would reflect the use of the DESCENT rule. After the failure, the trace would be "An increase in goal A leads to an increase in goal D".

The second comment to make about Fig. 6 is that more than just rules are being accumulated. Each time an input argument causes a failure, some heuristic is blamed, and each episode offers another clue to what is wrong with the heuristic for the given domain. The heuristic remains in use however until some generalization can be made about the kinds of situations where the heuristic doesn't work.

The serious use of generalization to construct replacement heuristics is a research topic for the future. The interested reader is referred to the large literature on induction, some pointers to which appeared at the beginning of this article. Currently, we are looking at the simpler problem of heuristic reordering.

In particular, consider the HIERARCHY heuristic "prefer goals of consumers". There are many cases where this heuristic fails to find the shortest path. For example,

Question:	If interest rates went up, what would happen to the federal budget deficit?
Novice Answer:	An increase in interest rates raises the deficit, because people can't buy things, which reduces business profits, which reduces tax revenues, which raises the deficit.
Better Answer:	An increase in interest rates raises the deficit, because the federal debt will cost more to finance, which raises the deficit.
Question:	If taxes were increased, what would happen to unemployment?
Novice Answer:	An increase in taxes raises unemployment, because people can't buy things, which reduces business profits, which means businesses have to cut back and lay people off.

Better Answer: An increase in taxes raises unemployment, because business
                        profits are reduced, which means businesses have to cut back
                        and lay people off.

Heuristic reordering means that when a heuristic fails several times, as the "prefer
the goals of consumers" does here, then the reasoner should try applying other existing
heuristics to see if they do better. The HILL-CLIMB heuristic, which prefers goals
sharing elements with the target goal, is a good candidate, because both of the better
answers link to goals with the same actor as the target goal.

Story understanding and economic reasoning

Novices, I have argued, tend to abandon goal-based reasoning in economics in favor
of a more neutral link-based search. The failure-driven mechanism of the previous
section would explain this as the result of repeated failures to apply goal-based
heuristics. The basic problem with novice reasoning is that it is very much like the
reasoning used in understanding stories or everyday life. Story understanding heuristics
track the goals of the main characters only (e.g. you don't normally worry about the
goals of the cook when reading a story about a romantic dinner engagement) (Wilensky,
1978). This works fine when the goals of each character are relatively autonomous.
The interesting parts of stories are often those places where goal interactions start
occurring and we are not sure what will happen.

Story understanding heuristics fail in economic reasoning not because economic
actors don't have goals, but because there are too many such goals and too many
interactions. Heuristics that minimize interactions will fail to see important possibilities.
For example, the heuristic "prefer the goals of consumers" seems reasonable if you're
only interested in how consumers are affected by changes in the economic situation.
Unfortunately, the heuristic not only generates weak arguments in situations where
the effects on consumers is largely irrelevant, but may even fail to find the most
important effects on consumers, by neglecting how a large effect on business translates
back into an effect on consumers.

Conclusion

We have tried to whittle down the problem of learning expertise to something feasible
for implementation, but still true to the spirit of cognitive modelling. Reasoning in
the economics domain offers a rich spectrum of levels of expertise. One level is the
literate novice, who understands most of the necessary concepts such as interest rates,
inflation, and deficit, but whose reasoning is based primarily on how changes affect
some favorite economic actor, e.g. business or consumer. A later level is the facile
debater of the newspaper editorial, who uses the same concepts as the novice, but
whose reasoning produces shorter, stronger arguments that take into account how all
the relevant parts of the economy might be affected. This reasoner is not a true expert,
but is certainly more expert than the novice.

The model of the novice has a knowledge base of highly interconnected goal forests.
The search heuristics must be very selective to avoid a combinatorial explosion in
search time. They are story understanding heuristics, focussing on favorite actors and
"moralistic" outcomes (e.g. good comes from good, bad from bad).

The model of the more expert reasoner has the novice knowledge base amended into two ways. First, good arguments have been collected for various issues and coded into explicit, specific rules for linking important economic quantities directly. Second, the story understanding heuristics have been replaced with more appropriate ones for the "amoral" multi-character realities of the economy.

The mechanism of this charge is a failure-driven process that not only records better arguments as they are seen, but also records the failure of existing inference rules to find these arguments on their own. The collected failures are used to determine which search heuristics work best in which situations.

Our hope is that this mechanism can be used to account for a significant part of the changes in reasoning style from novice to expert. Experimental research on the difference between novice and expert reasoning (e.g. Chi, Feltovich & Glaser, 1981; Johnson *et al.*, 1981) is compatible with our own informal observations that experts take into account more alternatives than novices, but that they have available explicit procedures for quickly choosing an alternative. In our model, an expert has learned specific argument chains, but these chains give access to the alternatives from which they were selected.

It is also our belief that this approach to modelling learning applies in general to many domains where expertise has what is often called a "shallow reasoning" component. Consider for example the domain of medical diagnosis. Although goal trees and graphs are pretty useless for modelling either novice or expert knowledge, more basic aspects of our approach to knowledge representation and reorganization seem quite portable.

First, novice knowledge is *domain-centred*, that is, it is organized around the basic physical objects and causalities of the given domain. In economics, novice knowledge was organized around the economic actors, their goals, and their actions. In medicine, the organization would be around the body's organs, its diseases, and their physical effects.

Second, expert knowledge is *task-centred*, that is, it is organized around the kinds of answers that the expert wants to find. In economics, expert knowledge was organized around economic quantities and interactions between them. In medicine, the organization would be around the symptoms of various diseases and their interactions.

Third, the expert's knowledge is comprehensible to the novice and examples of expert reasoning is a major source for novice learning. In economics, the novice understands expert arguments by going from economic quantities and links to the underlying events and causations. In medicine, the symptoms and their interactions can be understood by going to the underlying diseases and their effects on the organs.

Fourth, the novice's reasoning is failure-prone, and, hence, ripe for learning from failure. In economics, worrying about the motivations of economic actors produces answers, but usually not very good ones. In medicine, thinking at the organ and disease level is likewise fairly inefficient. In both domains, novice reasoning is weak in handling complicated interactions, and this is where the reorganization towards explicit rules is most important.

This is all just speculation at the moment. In artificial intelligence work on diagnosis, Chandrasekaran & Sanjay (1982) present very similar views when comparing deep versus compiled knowledge. But I do not have any protocol analyses of novice and expert medical diagnoses that would convince a good cognitive scientist that novices become experts in the ways just suggested.†

In fact, years of work lie ahead of us on many fronts just in the economics domain. The model itself is under continual development and implementation, especially in the area of extending its basic qualitative reasoning power. Furthermore, our analyses of Salter's protocols and of the various published texts of expert arguments have only scratched the surface. It will be some time before we can determine how appropriate our reasoning or learning models are for these data [see Sleeman & Smith (1981) for a discussion of the combinatoric problems involved building and verifying models in a much simpler domain].

Despite these difficulties, we believe that even these preliminary investigations into the automatic acquisition of reasoning style will have important consequences for knowledge-based inference systems.

The Yale Economics Learning Project has included the following people over the past few years, all of whom have contributed to the ideas discussed in this paper: Drew McDermott, Mark Burstein, Gregg Collins, Shoshana Hardt, Stanley Letovsky, and William Salter.

This work was funded in part by the Air Force Office of Scientific Research under contract F49620-82-K-0010.

References

BURSTEIN, M. H. (1981). Concept formation through the interaction of multiple models. In *Proceedings of the Third Annual Conference of the Cognitive Science Society*, pp. 271–273. Cognitive Science Society.

CARBONELL, J. G. (1979). Subjective understanding: computer models of belief systems. *Doctoral dissertion*, Yale University. *Research Report # 150.*

CHANDRASEKARAN, B. & SANJAY, M. (1982). Deep versus compiled knowledge approaches to diagnostic problem-solving. In *Proceedings of the AAAI-82*, pp. 349–354. American Association for Artificial Intelligence.

CHI, M. T. H., FELTOVICH, P. J. & GLASER, R. (1981). Categorization and representation of physics problems by experts and novices. *Cognitive Science*, **5**(2), 121–152.

DE KLEER, J. & BROWN, J. S. (1982). Foundations of envisioning. In *Proceedings of the AAAI-82*, pp. 434–437. American Association for Artificial Intelligence.

DUDA, R. O., HART, P. E., NILSSON, N. J. & SUTHERLAND, G. L. (1978). Semantic network representations in rule-based systems. In WATERMAN, D. A. & HAYES-ROTH, F., Eds, *Pattern-directed Inference Systems.* New York: Academic Press.

FORBUS, K. D. (1981). Qualitative reasoning about physical processes. In *Proceedings of the Seventh International Joint Conference on Artificial Intelligence*, pp. 326–330. The International Joint Conferences on Artificial Intelligence.

GENTNER, D. (1982). Structure-mapping: A theoretical framework for analogy and similarity. In *Proceedings of the Fourth Annual Conference of the Cognitive Science Society*, pp. 181–184. Cognitive Science Society.

JOHNSON, P. E., DURÁN, A. S., HASSEBROCK, F., MOLLER, J., PRIETULA, M., FELTOVICH, P. J. & SWANSON, D. B. (1981). Expertise and error in diagnostic reasoning. *Cognitive Science*, **5**(3), 235–283.

KOLODNER, J. L. (1980). Retrieval and organizational strategies in conceptual memory: A computer model. *Doctoral dissertation*, Yale University (November). *Research Report # 187.*

---

† The work reported in Johnson *et al.* (1981) is provocative but involves novices who actually already have a fair degree of expertise in diagnoses.

KUIPERS, B. (1982). Getting the envisionment right. In *Proceedings of the AAAI-82*, pp. 209–212. American Association for Artificial Intelligence.

LEBOWITZ, M. (1980). *Generalization and memory in an integrated understanding system.* *Doctoral dissertion*, Yale University (October). *Research Report # 186.*

MICHALSKI, R. S. (1981). An application of AI techniques to structuring objects into an optimal conceptual hierarchy. In *Proceedings of the Seventh International Joint Conference on Artificial Intelligence*, pp. 460–465. The International Joint Conferences on Artificial Intelligence.

MITCHELL, T. M. (1978). Version spaces: An approach to concept learning. *Doctoral dissertation*, Stanford University.

NILSSON, N. J. (1980). *Principles of Artificial Intelligence.* Palo Alto, California: Tioga Publishing Company.

NORMAN, D. A. & BOBROW, D. G. (1977). Descriptions: a basis for memory acquisition and retrieval. *Memo 7703*, Center for Human Information Processing, La Jolla, California.

RIESBECK, C. K. (1981). Failure-driven reminding for incremental learning. In *Proceedings of the Seventh International Joint Conference on Artificial Intelligence*, pp. 115–120. The International Joint Conferences on Artificial Intelligence.

SALTER, W. (1983). The structure and content of tacit theories of economics. *Doctoral dissertation*, Yale University.

SCHANK, R. C. (1982). *Dynamic Memory: A Theory of Learning in Computers and People.* Cambridge University Press.

SLEEMAN, D. H. & SMITH, M. J. (1981). Modelling students' problem solving. *Artificial Intelligence*, **16**(2), 171–187.

STEFIK, M. (1981). Planning with constraints (MOLGEN: Part 1). *Artificial Intelligence*, **16**(2), 111–139.

WILENSKY, R. (1978). Understanding goal-based stories. *Doctoral dissertation*, Yale University. *Research Report # 140.*

WINSTON, P. H. (1979). Learning by understanding analogies. *Memo 520*, M.I.T. Artificial Intelligence Laboratory.

On the application of rule-based techniques to the design of advice-giving systems

PETER JACKSON† AND PAUL LEFRERE

Institute of Educational Technology, Open University, Milton Keynes MK7 6AA, U.K.

This article attempts to assess how much is known about building systems whose advice actually benefits users. We review current approaches to the provision of on-line help, and suggest that the most promising are those which represent a user's intentions explicitly. Following this lead, we examine recent work on speech acts, planning and meta-level inference for clues as to how a user's inputs could be interpreted in the context of his current aims and activities. We conclude that the appropriate context of interpretation for an input is supplied by hypotheses concerning the current state of a user's plan. Next we suggest that the techniques developed in rule-based systems could be used to implement an advisor capable of generating and revising plan hypotheses, and outline what we consider to be the outstanding problems of control associated with such an implementation. Finally, we show how such a system might help a user to structure his activity, so that he can iterate towards his goal while avoiding common errors.

1. Introduction

The automatic generation of advice poses many of the problems associated with man–machine dialogue in an acute form. Simple approaches to command processing and database interrogation can afford to ignore such issues as why a user wishes to execute a particular command, or why a user wants a particular piece of information. However, even the most rudimentary advisor must take a user's goals into account, otherwise there is no guarantee that the advice given will be appropriate. For example, users do not always issue the right commands in the right order or ask the right questions, and an advisor must try to uncover the intentions underlying such inputs. Advice is appropriate only to the extent that it helps a user to derive and debug a plan of action for achieving his aims.

In the absence of a well-articulated theory of advice, we attempt to analyse various aspects of advice-giving, such as informing, explaining and recommending. A brief review is provided of current work on user-friendly interfaces to show how these issues are usually addressed. This is followed by a consideration of the role of pragmatics in man–machine communication with regard to such matters as interpreting user inputs and inferring the goals of an interaction. We conclude that in order to support planful behaviour, a program must have access to some representation of its own functionality, and be able to reason about the way in which it is being used. The rest of the paper tries to show how some of the techniques associated with rule-based systems could be used to implement an advisor capable of responding to queries and commands in a context-dependent manner.

† *Present address*: Department of Artificial Intelligence, University of Edinburgh, Hope Park Square, Meadow Lane, Edinburgh EH8 9NW, Scotland.

DEVELOPMENTS IN EXPERT SYSTEMS
ISBN 0-12-187580-6

We make no attempt here to deal with aspects of presentation, such as screen layout and help vocabulary, which affect the legibility and comprehensibility of advice. Neither do we concern ourselves with the technology of the interface, e.g. speech recognition versus keyboard input. Our attention is confined to (1) the kinds of reasoning that might be required to ensure that advice is relevant and (2) the kinds of representation that might support such reasoning. We argue that the logic of the interaction is the central issue, and that progress in user-friendly systems depends uon developing formal methods of dialogue design. Far from solving such problems, the introduction of more sophisticated channels of man–machine communication requires their solution.

2. Interactive advice on current systems

A user who is unfamiliar with a system should be able to find out how to communicate with it and which of his activities it will support. Typical sources of information include advisory services (Coombs & Alty, 1980), other users (Lang & Auld, 1982), printed documents (Sullivan & Chapanis, 1983) and computer dialogues (Martin, 1973; Sime & Coombs, 1983). In the latter case, the information may be made available in various ways, including menus [as in WordStar: (Naiman, 1982)], an "apropos" facility [as in EMACS: (Stallman, 1979)] for looking up commands in a dictionary, and access to an on-line manual [as in Franz Lisp: (Foderaro & Sklower, 1981)]. Each of these facilities is a variant of the "help key" approach to advice-giving, whereby pre-stored text is accessed and displayed to describe such things as the syntax and semantics of commands and the options available to a user whilst in a particular mode. The case for this kind of provision has been argued as follows (Kennedy, 1974, p. 311):

> A help key which gives a detailed description of a selected part of the system is essential if only to give confidence . . . . If the user can ask the computer what to do at any stage or explain how a particular function works, the process of familiarization proceeds fairly rapidly.

The help key strategy is susceptible to various refinements. For example, Gaines & Shaw (1983) recommend that if more than one level of description is required, messages should be organized according to brevity, and progressively revealed as requested, with brief memory aids being offered before more detailed expositions. This "query-in-depth" facility can itself be augmented by referring to a user's previous actions. For example, in the Berkeley MELVYL system (Klemperer, 1982), the degree of detail in the help offered depends upon whether the user has just made an error.

Help messages of this kind relate primarily to the descriptive aspect of advice. Technically, it is not difficult to display pre-stored textual or graphical descriptions of the component parts of a system, or to give an overview of how those parts are linked; what has been lacking is a theoretical basis for choosing one description over another (Jagodzinski, 1983). What the user needs to know about a system at any given time depends mostly upon his plans and purposes. Providing progressively more detailed descriptions may eventually provide the user with the information that he needs, but it also presents him with a search problem. While he is solving this problem, he may be effectively distracted from the main task.

Help messages are often poorly understood, perhaps because their authors cannot craw upon ". . . theories of defining, explaining, directing and so on" (B. N. Lewis & Cook, 1969). There is often no meta-commentary to tell the user how to interpret a given message, e.g. as information about a state, notification of an error, a prompt for action or the provision of feedback (Thomas & Carroll, 1981). Other common failings are messages which tell users things they already know, or which consist of little more than a list of options, none of which actually meets the user's needs.

Theoretical shortcomings are also apparent when it comes to explaining how a particular function works. In the absence of a theory of explanation, some form of simulation often substitutes, such as exploration or demonstration. For example, exploration can take the form of a "reconnoitre mode" in which users investigate the effect of choosing particular options (Jagodzinski, 1983). Alternatively, a simulation based tutorial facility can allow a user to "execute a standard procedure, single step it, or practice it while the system monitors him" (Stevens, Roberts & Stead, 1982, p. 9). Nevertheless, from the point of view of giving advice, such devices are little more than dynamic descriptions which show, instead of tell.

In the CADHELP system (Cullingford, Krueger, Selfridge & Bienkowski, 1982), a sophisticated demonstration facility has been incorporated into a computer-aided design program. Domain-specific knowledge about graphics commands is encoded in a conceptual dependency format (Schank, 1975), and includes information about common errors and how to recover from them. Pressing a help key gains access to a list of features that the system can describe, such as how to use a pointing device to select an object and drag it across the screen. The user can choose from three types of explanation: summary, normal and errors. "Summary" displays an explanation in fairly high-level terms, while "normal" is fully expanded and "errors" attempts to anticipate problems that users have with a feature.

CADHELP does not support mixed-initiative dialogue at present, and so it cannot monitor the user's behaviour or check that explanations are understood. However, it can use the underlying knowledge structure to generate prompts which become more and more laconic as the user gains experience during the course of an interaction. Further, the same representational scheme has been used to generate animation sequences which show the user how to perform physical actions, like manipulating the stylus (Neiman, 1982).

Demonstrations might serve a number of distinct but related functions in an advice-giving system. First, they could simply describe the features of a utility by showing them in action; this is obviously appropriate for an application like graphics, but perhaps less effective in less visual domains, such as file management. Second, they could explain how particular features work by examining the association between actions and changes of state in greater detail; this is essentially the role played by "causal chains" in CADHELP (Cullingford & Krueger, 1980). Third, they could be used to convince the user that verbal accounts of system function, whether on-line or in the manual, are in fact correct, and clear up any misconceptions that he might have. Thus, demonstrations can be used to inform, explain and convince, but they do not constitute advice unless coupled with recommendations concerning some preferred course of action.

Advising is more than describing or demonstrating, because it suggests both the existence of alternatives and the possibility of a preference on someone's part. Advising also suggests the existence of a problem, which may involve more than the mere lack of information, such as an error that must be corrected before the task can proceed. Thus advice must be based on more than a representation of system capabilities. In addition to having access to knowledge about the domain, an advice-giving program needs to be able to reason about the current state of the interaction. This in turn requires the ability to infer a user's goals from his inputs and determine what plan of action he is following.

Giving advice in the light of a user's actions is easiest where the number of possible aims is limited. For example, the DCL demonstration program (Shrager & Finin, 1982) recognizes five complete goals of users of the VMS operating system. Within those limitations, it can recognize common errors, and either volunteer advice or point to the existence of a manual or an on-line help message, for example (Shrager & Finin, 1982, p. 339):

(The user types)	COPY TEST.TXT EXP1.TXT
	DELETE TEST.TXT*
(DCL advises)	If you mean to be changing the name of
	TEST.TXT to EXP1.TXT you might have
	simply used the command:
	RENAME TEST.TXT EXP1.TXT
	The HELP command can tell you more
	about RENAME.

The problem-solving aspect of advice-giving is nicely demonstrated by the UNIX Consultant, "UC" (Wilensky, 1982), which has a natural-language interface supported by PANDORA, a sophisticated planning system (Faletti, 1982). By calling upon PANDORA, UC can deal with quite complicated requests, such as "how can I save this file temporarily if I have no space left and can't contact the system manager?" (answer: mail it to yourself).

PANDORA is a frame-based program which uses hierarchical planning and meta-planning. It is able to (1) detect goals in commonly occurring situations; (2) resolve conflicts between goals and (3) project the consequences of planned actions onto a future database. Meta-goals are treated just like ordinary goals, in that the normal planning process is applied to them.

This system shows one way in which commonsense knowledge about a domain can be combined with the dynamic creation, execution and revision of plans to provide advice in an event-driven rather than a menu-driven fashion. Advice is event-driven by virtue of a frames package which contains a stock of commonly occurring situations and watches for inputs that invoke particular frames. Associated with these frames are rules of inference which might apply when a given frame is active. Thus selection of the "correct" frame is crucial, both for detecting goals and for generating plans. Selecting a frame by some partial matching operation is relatively easy when the number of candidates is small, but as the complexity and inter-relatedness of a frame system increases, difficult decisions have to be made about which parts of a frame are allowed to invoke the whole, and how to choose between alternatives when more than one frame matches. Faletti (1982, p. 186) admits that some of these decisions have yet to be taken, but points out that the connections between situations and goals has been obvious so far in the domains to which PANDORA has been applied.

In this review, we have tried to distinguish the descriptive and explanatory aspects of on-line help from advisory aspects involving recommendation and problem-solving. There is no doubt that informing and educating a user may be a necessary prelude to advising him; otherwise, recommendations may not be understood or accepted.

Nevertheless, description and explanation do not in themselves constitute advice, particularly when there is no feedback from the user concerning his understanding or acceptance of the assistance that is being offered.

Therefore we argue that the plan-based approach to providing on-line advice is the one with the greatest promise, and that it is capable of assimilating what has been learned about description and demonstration. However, we would like to explore an alternative formulation of the problem, based not upon scripts or frames but upon the techniques associated with rule-based systems. It is not the purpose of this exploration to re-enact the debate between proponents of higher-level knowledge structures, such as scripts and frames, and proponents of more formal methods, such as production rules and predicate logic. However, rule-based systems have successfully represented quite large and complex bodies of knowledge, and a considerable amount of work has already been done on pattern-directed rule invocation and conflict resolution. Consequently, it makes sense to ask how far some of these techniques could be applied to the management of man–machine interfaces, since it is possible that what an expert knows about a computing system is more naturally represented as a relatively large number of situation-specific rules than as a library of scripts in which goals and methods are to some extent pre-packaged. Script- and frame-based programs seem to work best in commonsense domains where behaviour is circumscribed by a comparatively small number of stereotypical situations, e.g. eating in a restaurant and travelling by air. While this may adequately characterize the interaction with some system utilities, such as filers and database packages, it is probably less true of others. Consider a programming environment, such as that available on the Lisp Machine (Moon, Stallman & Weinraub, 1983), involving numerous facilities including an interpreter, debugger and editor, where the user has a great deal of freedom as regards the way in which he approaches his task. It could be that scripts are not the best way to cover the space of behavioural possibilities associated with such a system, and that it might be better to detect goals and generate plans by composing a looser aggregate of propositional or procedural elements, even if this poses considerable problems of control. Also, the fact that rule-based systems have already been used to represent and reason about domains of realistic complexity renders them attractive as a basis for constructing advisors for such domains. This is not to say that existing expert systems should be reconfigured to serve as advisors; the experience with adapting such systems for tutorial purposes suggests that this would be a non-trivial task (Clancey & Letsinger, 1982). Rather, we are suggesting that extending the capabilities of rule-based systems beyond that of problem-solving, to include advisory and tutorial functions, will benefit knowledge-based systems in general, as well as providing another way of looking at the problems of the interface. Ideally, one would like a rule-based program to be able to use the same representation of knowledge in different ways, e.g. to solve problems, provide explanations, justify its decisions, generate examples and offer advice. An intelligent interface which served as advisor and problem-solving monitor using the same set of rules would be an application of both theoretical interest and practical utility.

In summary, users performing well-specified tasks such as data entry can clearly benefit from the provision of help that is purely descriptive. The number of alternative courses of action available at any one time is strictly limited, and the user has very little freedom as regards the factoring of the main task into subtasks and the way in which he iterates towards his goal. Where fuller explanation is required, demonstration may profitably be used to augment description, and the user may benefit from being allowed to explore the system and experiment with it under the supervision of a tutorial program. However, more open-ended tasks require more than information and explanation; the user needs advice about how to proceed. Further, advice should be on-line, dealing with real problems as they arise, rather than being confined to an off-line phase of exploration or demonstration.

The principal problems facing an advisor for open-ended tasks appear to be the following:

establishing what a user is trying to do;
helping a user to plan ways to reach his goals;
allowing for a two-way meta-commentary;
giving feedback on the progress or side-effects of commands and deciding when to volunteer advice or to question the user.

In the following section, we shall argue that existing emphases upon natural language understanding, input–output technology and domain-specific knowledge (e.g. Tanaka, Chiba, Kidode, Tamura & Kodera, 1982) must be augmented by a high-level and largely domain-free view of the way in which advisory dialogues can and should be managed. Regardless of the physical aspects of the interaction, the symbolic aspect remains the same: an intelligent system is faced with the problem of divining a user's intentions from his inputs. It is suggested that formal approaches to interpretation, planning and inference can advance the theory of man–machine interaction in this respect, and facilitate the application of artificial intelligence techniques to the interface.

3. Speech acts, plans and meta-level inference in man–machine dialogue

A man–machine dialogue can be considered as a manifestation of a physical symbol system, in the sense of Newell & Simon (1976). One way of talking about such systems is to talk about the language in which the exchange is conducted. Symbol structures can be used to perform a variety of speech acts, such as informing, requesting and ordering. Such behaviours are a potential source of information regarding the goals of an interaction and the plans that participants are pursuing. The difficulty is to describe formally the relationship between the symbolic exchange and these underlying intentions.

3.1. INTERPRETING USER INPUTS

Formal languages used for making statements, asking questions and issuing commands can be considered as pragmatic languages along the following lines. The sentences of such languages are syntactic entities whose physical manifestations as utterances or inscriptions invite the addressee to a behavioural response to the semantic valuation of their propositional content. The evaluation scheme employed and the response elicited depend upon pragmatic aspects of the interaction, such as the mood of the sentence and the context of the physical manifestation.

An example will make this clear. Suppose that a user wants to open a file called TEXT.DOC. Then the inscriptions "Is TEXT.DOC open?" and "Open TEXT.DOC!" share a common propositional component (that TEXT.DOC be open), and differ only in their mood (interrogative versus imperative) which suggests a different evaluation scheme and action in each case. The question implies that the user does not know whether the proposition is true, and requires the system to supply this information, while the command presupposes that the proposition is false and requires the system to perform some action that will make it true.

The difference between a language defined in terms of sentence tokens as opposed to a language defined solely in terms of sentence types has been characterized by Cresswell (1973) as the difference between an utterance language and an abstract language. It is clear that sentence tokens such as utterances can have different occasions of use which may constitute different contexts of interpretation, while sentence types cannot. The term "pragmatics" has been used in connection with this distinction to indicate variously: the relationship between signs and interpreters, involving the uses and effects of signs in some behavioural context (Morris, 1971); and that part of language which is concerned with the interpretation of indexical expressions (Bar-Hillel, 1954; Montague, 1972).

More recently, there has been a call for the integration of semantics and pragmatics with a general theory of cognition (Kasher, 1977; Bartsch, 1979). Attempts to explicate "pragmatical correctness' in terms of the appropriateness of expressions with respect to situations and goals are suggestive of the kinds of competence we would like to foster on both sides of the man–machine interface. Indeed, "pragmatical competence" in a language has been defined as the ability to handle the appropriateness relation between the expressions of that language and interactional goals.

The kind of analysis adopted here draws upon these intuitions, together with related ideas in the work of Austin (1962), Searle (1969) and D. K. Lewis (1972). It makes explicit the separation between (1) propositional acts, which perform the essentially semantic functions of predicating and referring, (2) illocutionary acts such as asking or commanding, which reveal a speaker's intentions, and (3) perlocutionary acts, which elicit some verbal or non-verbal response, such as compliance or a helpful reply.

We argue that successful trade in sentence tokens depends upon at least three layers of disambiguating context: (1) the illocutionary context supplied by the current goal of the interaction (e.g. what are we opening the file for?); (2) the perlocutionary context supplied by some plan for achieving it (e.g. getting into the right mode and typing commands that will elicit the right response from the system and have the desired effect) and (3) the propositional context supplied by the entities in the universe of discourse (e.g. which file are we talking about?). Even the syntax of simple command and query languages permits ambiguity in each of these respects. Intentions can be ill-formed (e.g. trying to write to a read-only file); plans can fail (e.g. trying to delete a protected file without reprotecting) or have unwanted side effects (e.g. deleting wanted files by using "wild" characters too liberally); while file specifications can be ambiguous with respect to different extensions, devices and so on.

Cohen & Perrault (1979) argue that a plan-based theory of speech acts should specify at least the following: (1) a planning system, i.e. formal languages for describing such things as the state of the world, legal operators for changing the state of the world, well-formed plans, etc. and (2) definitions of speech acts as operators in such a system, together with their effects and applicability conditions. The formal languages of (1) clearly require both a syntax and a semantics, whereas the definition of speech acts as operators upon formal representations supplies a pragmatics for the utterance language in which communication is conducted. It is the role of these operators to make explicit the relationship between tokens of this language in some context of use and the plans of speakers and listeners in that context. Once this relationship has been rendered perspicuous, a program has some basis for inferring a user's needs from a combination of expressed intentions (what he says he wants to do), overt actions (what he actually does) and explanations (what he says about what he does). It may be that no single source of evidence is sufficient for the majority of circumstances, but that taken together these sources may serve to guide the generation and refinement of hypotheses concerning the goals of the interaction.

Accordingly, we shall argue that pragmatics at the interface has three main aspects. First, the user could use a formal language to inform the system of his intentions and request information, as well as simply issuing commands. Second, the system could infer and refine a user's intentions from such inputs, by exploring the conditions and consequences associated with their propositional contents. Third, each party could employ a meta-language to discuss and debug the interaction itself, whenever there appears to be a mismatch between the expectations of system and user. In order to explicate the notion of expectation in this context, we shall need to consider the role of plans in more detail.

3.2. SUPPORTING PLANFUL BEHAVIOUR

It is useful to think of a physical symbol system as a realization of an abstract problem-solving exercise involving certain choices to do with the recursive factoring of goals into subgoals in the performance of some task.

A plan for achieving some goal is essentially a specialization derived by instantiation from a plan schema implicitly supported by the system. What the user does in performing the task, such as getting into the right mode and typing the correct commands, corresponds to an execution of this plan. For example, a word processor offers general facilities for tasks such as creating files and saving them on disk, while an individual user in carrying out his plan will create a particular file of a certain size, and try to save it on a particular disk with only a certain amount of space left on it. The user's plan may be well-formed and properly executed and yet fail, because the original intention, e.g. to create a large file and save it on a nearly full disk, was ill-conceived.

Amongst other things, an intelligent system would try to ensure that its plan schemas were lawfully instantiated and that the resultant plans did in fact succeed. The user and the system would need to agree about which plan schema the curent plan was instantiating, and what the consequences of plan execution would be. Plans with negative outcomes would need to be recognized and revised, while the user should be encouraged to generalize plans with positive outcomes and avoid common sources of error and inefficiency.

Early work on planning (e.g. Fikes & Nilsson, 1971; Sacerdoti, 1974) was mostly concerned with plan formation (deriving a sequence of operators for achieving a goal) and plan generalization (deriving a plan schema from a successful plan). While much remains to be done in these areas, more recently (e.g. Schmidt, Sridharan & Goodson, 1978; Genesereth, 1982) attention has been focused upon plan revision (deriving a new plan when the current one fails) and plan recognition (inferring a person's plan from his behaviour). This is due in part to the realization that the problems of representation and control associated with planning impinge upon crucial aspects of natural language understanding and belief organization (e.g. Schank & Abelson, 1977).

Each of these four aspects of planning is relevant to man–machine interaction. An intelligent system should be able to (1) provide planning aids for the user at the formation stage, (2) recognize the user's plan as an instantiation of one of its plan schemas, (3) revise its own hypothesis about the user's plan (should it turn out to be false) and help the user to revise his plan (should it fail) and (4) promote the generalization of plans which succeed. These desiderata are easy to state, but difficult to engineer.

Traditional approaches to the provision of aid during plan formation fall into two general categories: structuring devices, such as prompts and menus, which present the user with a number of options in his approach to the problem, and information facilities, such as command dictionaries and on-line documentation, which provide help without ordering the user's activity. This division is not absolute, of course; some menus serve both functions by informing the user as well as eliciting choices. Work on structured programming environments (e.g. Miller, 1982), goes a step further by introducing a vocabulary of concepts about planning and debugging which attempts to make various design options explicit. Such systems must be able to engage in some kind of dialogue with the user about his intentions and preferences.

The intentional aspect of planful behaviour is also central to plan recognition. Contemporary approaches to question-answering (e.g. Allen, 1983) based on computational models of speech act theory (e.g. Cohen & Perrault, 1979) emphasize the importance of inferring a speaker's underlying goals in generating a helpful response to a given query. Also, work on intelligent teaching systems suggests that the provision of appropriate advice requires the ability to entertain hypotheses about a student's problem-solving strategy.

Plan revision interacts with plan formation and recognition in complex ways. First, the system's hypothesis about a user's plan may be proved to be wrong in the light of subsequent inputs, so that its representation of that plan must be revised. Second, the initial hypothesis may be correct, but the user may spontaneously decide to change his plan, or revert to an earlier plan. Third, the user's current plan may actually fail, in which case the system should be able to give advice about alternative ways of achieving the goal. In general, the system is faced with difficult decisions concerning where to allocate blame when its expectations concerning inputs are not fulfilled, and whether to intervene or not.

Plan generalization is a form of inductive learning, and ideally one would like both components of a man–machine system to learn from their experience with past interactions. For example, the user should learn which kinds of plan are supported by the system, while the program should learn to anticipate likely planning errors. There is also the question of promoting optimal strategies and discouraging grossly inefficient ways of doing things.

3.3. REASONING ABOUT CONTROL

The relationship between planning and the interpretation of user inputs outlined above evidently requires a program to be able to reason about its own behaviour while it is being run. The distinction between that part of a program which deals with the execution of some task and that part which deals with planning the task is typically drawn in terms of the difference between base- and meta-level architecture (Genesereth, 1983). The idea is that the base-level interpreter accesses and manipulates the representation of the universe of discourse, while the meta-level interpreter determines which base-level actions should be performed and in what sequence.

Reasoning about knowledge is relevant to a number of problems in the design of intelligent systems, including explanation, conflict resolution and knowledge base revision (Lenat *et al.*, 1981). From the point of view of explanations which justify system decisions, it is generally assumed that a recapitulation of program actions is sufficient, so long as the correct level of description and detail is chosen [as in the TEIRESIAS knowledge transference system (Davis, 1983)]. However, it has been suggested that explanations for pedagogical purposes require a good deal more than this, since significant aspects of domain expertise are typically compiled into the specification of the rules in the knowledge base, as Clancey (1981, 1982) pointed out during the development of the GUIDON teaching system. That is to say, strategies concerning which rule to apply and the best order for expanding subgoals are implicit in the ordering of rules in the rule set and the ordering of conditions within rules.

Several desiderata emerged from the GUIDON and TEIRESIAS projects concerning knowledge representation in expert systems which incorporate aids for the transference of expertise from human to program or vice versa. First, knowledge in programs should be explicit and accessible. Thus, the criteria for ordering and structuring information of various kinds should be made clear and the consequences should be well understood. Second, programs should be given access to an understanding of these representations. One of the major benefits that might accrue from this policy is the ability to put a single representation of knowledge to different uses. As Davis & Lenat (1982) point out, rules can be viewed as code to be executed, or data structures to be examined to produce explanations.

It is clear that many researchers are approaching these problems from different perspectives which are nevertheless complementary, and which may eventually furnish designers with some basic principles and techniques for the implementation of more helpful systems. In the next section, we sketch an intelligent support system for editing tasks, which are more open-ended than data entry or file manipulation. Editing is an open-ended task to the extent that the user has freedom with regard to such matters as the organization of low-level actions and the degree to which iterative refinements are possible. We suggest that an intelligent support system should allow a user to concentrate on what he wants to do rather than on how he is going to do it on a particular system.

4. Context-dependent interpretation of questions and commands

The following sections explore the possibilities of using the techniques associated with rule-based systems to provide context-dependent advice. Although there exists an extensive literature on the logic of interrogatives [see the bibliography in Belnap & Steele (1976)], it is not easy to see how the various formalisms on offer might help with the management of a man–machine interface. This is partly because of the issue of context-dependency, and partly because most models of question-answering are more inclined towards database interrogation than requests for help or explanation. The literature on imperatives seems to be somewhat smaller and less distinct from the general literature on such related matters as performatives, deontic logics and algorithms. A thorough treatment of the main problems facing a logic of commands can be found in, for example, Rescher (1966).

4.1. REPRESENTING THE DOMAIN

One can use the formal languages associated with rule-based systems to try to describe what an expert knows about a domain like copy-editing. The syntax of such languages could conform to that of production rules, say, or to the Horn clause subset of logic. The expressive scheme which is actually chosen is perhaps less important than the ways in which the resultant representations can be interpreted.

For example, a production rule for opening a file in WordStar might be stated in terms of the following conditions and actions:

IF   you are at the No-File menu
        and you type D
        followed by a file specification
        and that specification is legal
THEN   WordStar opens a file of that specification

while an equivalent Horn clause representation might give the action as a conclusion followed by a list of conditions:

WordStar opens a file with a particular specification
IF   you are at the No-File menu
     and you type D
     followed by the file specification
     and that specification is legal.

We have tried to make the English rendering of such rules neutral with regard to the ways in which they might be applied. Under one interpretation, they can be seen as recipes, or plan generators: if you want to achieve this, then do that. Under another, they can be construed as explanations, or trace generators: the system did this because you did that. There need be nothing either in the formalism or in the executable code, be it stylized Lisp or Prolog clauses, which determines how the rule will be used. The advantages of using representations of knowledge which are sufficiently flexible to be put to a variety of uses has been stressed by both advocates of production rules (e.g. Davis & Lenat, 1982) and advocates of logic programming (e.g. Kowalski, 1982).

One of the most obvious uses of rules which specify preconditions is to represent the factoring of goals into subgoals for complex tasks. An example would be rearranging blocks of text while copy-editing. Thus a "cut and paste" operation in WordStar might be represented by the following set of rules:

WordStar cut and pastes a block of text
IF   you mark the block
     and move the block
     and tidy the block

WordStar marks a block of text
IF   you mark the beginning of the block
     and mark the end of the block

WordStar marks the beginning of a block
IF   you put the cursor at the beginning of the block
      and type CONTROL-K
      followed by B

WordStar marks the end of a block
IF   you put the cursor at the end of the block
      and type CONTROL-K
      followed by K

WordStar moves a block to a new position
IF   you put the cursor at that position
      and type CONTROL-K
      followed by V

WordStar tidies a block
IF   you unmark the block
      and reformat the block

WordSar unmarks a block
IF   you type CONTROL-K
      followed by H

WordStar reformats a block
IF   you put the cursor at the start of the block
      and type CONTROL-B

The rules given above specify preconditions of various states of affairs than the user might want to bring about. However, these states might also have consequences or "postconditions" which follow whenever they are actualized. For example, opening an extant file generally results in the contents of that file being copied into the edit buffer or temporary file, while saving buffer contents usually results in the new file being accorded a certain default protection and the backup of the file being updated automatically. It is unlikely that a user would ever view either of such outcomes as specific goals, representative of his intentions. Rather, they are a facet of the way the system works, and he can only change these conventions to the extent that the system is customizable. If the user does not know how the system works, then many of his actions will have unforeseen consequences, not all of which may be desirable.

Thus, conclusions such as

WordStar saves an edit of a file
IF   you type CONTROL-K
      followed by S

which are specified in terms of their conditions, can be viewed as conditions of conclusions which are their consequences in rules like

WordStar updates the backup of a file
IF   WordStar saves an edit of the file.

We shall assume a knowledge base, K, consisting of such statements, or their production rule equivalents, together with a set of propositions, S, which indicate the status of system parameters. In a logic program, these might be represented as ground literals in the database, i.e. atomic propositions or their negations which contain no variables. In a production system, they might be stored as attribute-value lists in working memory.

Rules which specify the consequences of commands can be used to manipulate a model of the current state of the system. Amongst other things, this allows the meta-level to reason about the side effects of performing actions in response to user inputs. For example, the user may decide that some of the unforeseen effects of his commands are undesirable, and may wish to know how they came about and whether they can be undone.

Of course, there are certain difficulties associated with the maintenance of consistency and completeness in modelling dynamic systems of any complexity. Most of these difficulties relate to the "frame problem" formulated by McCarthy & Hayes (1969), i.e. the problem of deciding which old assertions no longer hold true after a change of state and which new assertions must be added. We shall do no more than mention the problem here, while noting that various solutions have been proposed (e.g. Hewitt, 1969; Fikes & Nilsson, 1971; Kowalski, 1979, Ch. 6).

4.2. REPRESENTING THE INTERACTION

Sentence types will be represented by the concatenation of a proposition and a mood symbol from the set [?, !]. If p is an atomic proposition, then let the informal reading of "p?" be "how do I make p true?" or "how do I do p?", and let the informal reading of "p!" be "make p true!" or "do p!". Thus, the common propositional content participates in the different illocutionary acts of querying and commanding. (For the moment, we confine our attention to "how" questions, as opposed to "yes–no" or "wh–" questions.) In their perlocutionary aspect, the question expects a helpful response, while the command expects compliance, since "how do I do p?" implies "I want you to tell me how to do p" and "I think you know how to do p", while "do p!" implies "I want you to do p" and "I think you can do p".

Sentence tokens are physical manifestations of sentence types in some context of use which specifies both the purpose of the interaction, in terms of goals and plans for achieving them, and the domain of interpretation, in terms of the universe of discourse. The current state of this universe will be given by atoms in S, while operators for changing this state are given by the rules in K. We shall assume that both kinds of information can be accessed and manipulated by the base-level interpreter in response to questions and commands.

Information about the current state of the interaction, on the other hand, can be represented by a plan, or plan hypothesis, which may be only partially developed and instantiated. This information is not represented in K or S, but in a some structure for representing goals, such as a stack or tree or net. It is not available to the base level of the interpreter, but to a meta level which tries to interpret inputs in the light of the current state of the plan.

4.2.1. Aspects of plan formation
The best way to illustrate the way that meta-level inference might proceed in a simple advice-giving system is by example. Suppose that a user wishes to open a file but does not know the correct commands to type or in what context to type them. Assume, for

the sake of argument, that the user knows, or can easily find out, how to obtain this information by typing a query, for example

HOW OPEN TEXT.DOC?

The propositional content of the query is then set up as a goal. The basic evaluation scheme employed is that of backward reasoning: given the conclusion that we wish to establish, e.g. that TEXT.DOC be an open file, what conditions would have to be satisfied? Given a rule for opening files in K like the one cited above, the answer is provided by the instantiation

WordStar opens TEXT.DOC
IF   you are at the No-File lmenu
      and you type D
      followed by TEXT.DOC
      and TEXT.DOC is a legal specification.

There could be a further rule which defines what constitutes a legal specification, and so judgement could be pronounced as to whether the final condition in the instantiation has been satisfied by the application of this rule. In general, some or all of the conditions of a rule may be capable of evaluation by the application of other rules, and the recursive replacement of such conditions by their own preconditions constitutes a form of plan expansion. The level to which the plan needs to be developed will depend upon the individual user. For example, in the cut and paste operation described above, the user may or may not know how to perform subtasks like marking and reformatting.

Unrestrained, the plan expansion process should bottom out with a fully specified plan for achieving the main goal, assuming that the rule set is complete in this respect. Controlling the level of detail in plan description therefore corresponds to controlling the deduction, which is clearly an issue for the meta-level of the interpreter. This problem relates to points raised elsewhere concerning the appropriate level of detail in the presentation of proofs (e.g. Chester, 1976; Sergot, 1983), and in the recapitulation of program actions (Davis & Lenat, 1982).

In the example given above, the deduction will usually fail to establish the goal. That is to say, TEXT.DOC will presumably not be open at the time of the query, since the user's intention is to bring that state of affairs about. If the file is already open, then the user should be apprised of this fact when the deduction succeeds. Thus it is clear that "yes–no" questions could be answered by the same process of backward reasoning, possibly augmented by conventions concerning negation as failure, while "wh–" questions could receive a similar treatment, using the standard techniques of answer extraction. The principal use of such questions might be to assess the state of system parameters, such as flags and switches, or find out what mode one was in.

As well as generating a plan, the expansion of goals into subgoals could also go some way towards creating a context of interpretation for subsequent inputs. The illocutionary context will be given by a statement of intention supplied by the user; the perlocutionary context will be given by the current state of the plan in terms of subgoals satisfied so far and subgoals pending while the propositional context will be given by referents in the query, such as files and devices named by the user. If no plan is forthcoming, then the goal does not instantiate one of the goal schemas supported by the system, and this constitutes a failure of illocutionary context. An example would be trying to "undelete" a file when such an operation is not supported by the system. If a plan is forthcoming, but turns out to be unworkable or to have unwanted side effects, then this is a failure of perlocutionary context, since the desired response is not forthcoming. An example would be trying to delete a protected file, or expunging a directory to make space and then realizing that one should have archived the deleted files first. Finally, the plan may be well-formed but the entities involved may no longer exist; this is a failure of propositional context. An example would be trying to print a file which has been deleted. Illocutionary failures are not recoverable within the system, so the original intentions must be reformed. If you cannot undelete a file, then you must retype it or check with the system manager to see if there is a dumped copy. Perlocutionary failures require the formation of an alternative plan; if no such plan is forthcoming, then the intention must be reformed. Thus one may be able to reprotect a file one wishes to delete, but one is unlikely to be able to restore files which have been purged. Propositional failures may or may not be recoverable, depending upon whether the lost context can be restored, e.g. by undeleting files.

In addition to displaying the conditions associated with a particular goal, it might also be helpful to inform the user of the consequences of achieving it. It could be that success would have certain unwanted side effects of which the user is unaware. For example, editors which use "kill" and "yank" to copy blocks of text frustrate users of all levels of expertise from time to time. Incidental deletions and insertions which were not part of the original plan frequently displace the killed text from the buffer, whence it is usually gone for good, Overwriting the buffer is an unwanted side effect of an otherwise sensible action.

Rules which inform the user of the consequences of success would have to be executed bottom-up. That is, the goal statement would match with the conditions of such rules, rather than their conclusions, as is the case in top-down execution. Consequences would not become actual until all of their conditions were satisfied, and so the user need not be alerted until the conclusion is imminent. Also, meta-rules could be used to prevent the user from being burdened with useless information. For example, we need only warn him about the kill buffer overwrite if the buffer contains a large block of text which has not been yanked yet.

The plan formation process described above is top-down and hierarchical, since it proceeds via the orderly expansion of goals into subgoals. This is obviously not the only way to form plans, but it is the simplest to expound and will suit our present purposes. Accounts of more opportunistic and heterarchical approaches to planning can be found elsewhere (e.g. Hayes-Roth & Hayes-Roth, 1979).

4.2.2. Aspects of plan revision
Once a system has some representation of a user's plan, it has some basis for interpreting his subsequent behaviour, criticising it and providing advice (London & Clancey, 1982). In the example given above, the search for a solution to the user's problem is initially top-down and model-driven. It has been suggested that such searches work best in an area where the number of possible solutions is small (Genesereth, 1982). This is mostly true of domains like copy-editing and elementary programming, but less plausible in domains like medical diagnosis and geological consultancy. On the other hand, bottom-up search spaces generally lead to combinatorial explosion, unless curbed by search

heuristics (Kowalski, 1979). In some cases, top-down and bottom-up search can be combined, particularly if one has explicit knowledge of the user's goal (Genesereth, 1978).

Given that the meta-level of the interpreter has derived a plan hypothesis from its evaluation of the query, subsequent commands from the user can be evaluated bottom-up, i.e. by matching their propositional contents against unsatisfied conditions in the plan. As such conditions are satisfied, the proof which failed top-down approaches success bottom-up. In a perfect world, the hypothesis is correct and the user's actions instantiate the remaining parts of the plan.

The plan hypothesis is no more than a prediction, however. Such predictions are prescriptive, because they represent what the user should be doing, given a statement of intent and the representation of the rules in K. However, the system's problems are not over, because the user may (1) change his goal; (2) adopt some other strategy for achieving the same goal; (3) interpolate another task or (4) type the wrong command or type commands in the wrong order. Deciding which of these four possibilities has occurred when expectations are violated is unlikely to be easy.

In the event of (1), one might expect some explicit statement to this effect from the user, a speech act which refers to the interaction itself, rather than the task domain. Some justification for the change of plan might also help the system suggest alternatives. For example, the user might not like one of the consequences of achieving his goal and therefore need assistance in reforming his original intention.

Faced with (2), the system should be able to backtrack and realize that although the user's input is not relevant to the plan as developed by the system, it is relevant to an alternative plan which the system is able to support. In general, there appear to be two possible solutions to this problem. One can generate all possible plans (to some level of detail) when faced with a goal statement to yield multiple expectations. This is the course taken by London & Clancey (1982) in their IMAGE module of GUIDON2. Alternatively, one can generate a single plan, selected from those on offer according to some criterion, and then revise it if expectations are violated. This is closer to the philosophy of the BELIEVER system (Schmidt *et al.*, 1978) which employs "plan critics" after each stage of plan expansion, along the lines of Sacerdoti (1975). Which option is better depends upon the nature of the underlying search space. In the editing domain, it may not be feasible to generate all possible plans for many of the goals that users wish to achieve.

To deal with (3), the system would have to be able to suspend and reactivate plans, perhaps by pushing and popping whole contexts. It is easy to see that this could pose all kinds of problems and quickly get out of control. (4) requires the system to recognize the input as a faulty execution of a plan, rather than a rejection of the plan or a change of intention, and perhaps bring to bear specialized support software, such as a spelling corrector. Deciding what category of "action slip" (Norman, 1981) had occurred would be extremely difficult in general.

No matter how sophisticated one's interface program, it is unlikely to perform well when expectations are violated unless there is some meta-communication between user and system. Thus the system needs to be able to ask questions, such as "is that what you intended?", "do you still want to do *p*?", "shall we shelve this plan or scrap it?" and so on. The answers to such questions will relate directly to meta-level inference concerning the dialogue, rather than to inference at the base level about the domain.

Sergot's (1983) "query the user" facility for logic programming provides a useful pointer in this respect. In this system, the interpreter is able to prompt the user for the information that it needs to solve a particular problem. The answers to such questions can then be used to perform base-level inferences and complete refutation proofs which have failed for lack of information. It should also be possible to ask questions about the user's goals and use these to perform inferences at the meta-level. The rules for doing this would be meta-rules concerned with the conduct of the dialogue, but they could be expressed in the same representation language as rules concerned with the domain.

4.2.3. Aspects of plan recognition

Plan formation as described above requires an initial query on the part of the user, which the program interprets as a goal statement. Thus meta-level inference is initiated top-down, with the goal matching against conclusions. However, we might also like our intelligent system to be able to entertain hypotheses concerning a user's plan in the absence of such a query or statement of intention, but simply as a function of parsing the stream of commands according to the rules in K, as if these rules represented a grammar for the interaction. In this case, meta-level inference would be initiated bottom-up, with the propositional contents of commands being matched against conditions. Plan recognition programs as currently implemented do not perform this kind of hypothesis generation "from scratch" with judgements concerning the appropriateness of user behaviour versus the appropriateness of the hypothesized plan.

There are many difficulties associated with doing this. The support program would have to decide at what point it has enough information to derive a hypothesis, or set of hypotheses. It would then have to evaluate these hypotheses using some kind of "group and differentiate" strategy in the light of further inputs. This process would be complicated by any slips of action that the user makes, so the methods employed would have to be sufficiently robust to cope with a certain level of noise. It would also have to decide at what point to intervene, if the user's input sequence did not generate any hypotheses. In general, the plan recognition process would almost certainly benefit from the ability to interrogate the user when it gets into difficulties over goal inference.

The program could be said to have arrived at a goal hypothesis when it recognized that the user's inputs satisfy some or all of the conditions of a particular rule. The conclusion of this rule could then be entertained as a hypothesis of the user's immediate goal, and the associated conditions could be used as a context of interpretation for the evaluation of subsequent inputs, as before. Violation of expectations would require plan revision along the lines mentioned earlier.

Plan recognition would have to proceed recursively, since the goal initially inferred by the program might be no more than a subgoal in some greater plan. Thus, when an inferred goal is achieved, its propositional content should be evaluated bottom-up, to see if it matches against the one of the conditions of another rule, with a view to entertaining the conclusion of this rule as an inferred, higher level, goal. Just as top-down inference should bottom out at some point as we run out of rules for expanding subgoals, so bottom-up execution should reach a ceiling when we run out of rules whose conclusions match the conditions of other rules, assuming that there is no circularity in the specification of the knowledge base.

For example, suppose that a user sets out to cut and paste a block of text. The rules given above for doing this can be viewed as a problem reduction tree, and top-down plan expansion typically involves visiting the nodes of this tree in preorder, i.e. starting at the root of the tree and considering successors depth-first and from left to right. Bottom-up evaluation of user inputs, given the correct plan, involves visiting these nodes in postorder, i.e. the inputs correspond to terminal nodes of the tree whose left to right visitation enables the visitation of the parent node.

Plan recognition in this context involves (1) inferring as soon as possible that the immediate goal is to mark the text; (2) inferring what the user intends to do with the marked block (e.g. rather than moving it, he might want to delete it, or write it to another file) and (3) monitoring the execution of the hypothesized plan. This process requires the program to mix forward and backward reasoning, and perhaps make guesses that may later turn out to be wrong. The generation of goal hypotheses is data-driven, but once a goal has been inferred, the generation of plan hypotheses is model-driven. Further data may require both plan and goal revision, and hence a repetition of this cycle. In any case, as subgoals such as marking are achieved, the program will have to go through the goal inference process again, with the successful subgoal as data.

Such deductions are not easy to control, and are unlikely to be very robust in the face of noise unless tightly constrained. A related problem is that users typically interrupt one activity to start or complete another, e.g. after marking a block to be moved, one might decide to alter the screen format or read a block from another file. Also many incidental cursor motions and insertions might be difficult to interpret, even if the current plan hypothesis were correct.

4.2.4. Aspects of plan generalization
In editing, there are a number of related problems to do with encouraging a user to abstract general methods from the performance of particular tasks. A user's responses to commonly occurring situations involving insertion and deletion soon become automatic, in the sense that after a while there is very little overt planning involved. Such rules as have been learned are often "macro-processed" for stereotypical tasks; that is to say, the small number of rule applications required to perform a simple task, like going back and deleting a word in the current sentence and replacing it with another word, are composed into a single action sequence which is triggered and executed as a whole.

This style of information processing aids fluency but often leads to inefficiency. Given some more complex task involving replacement or rearrangement, users often pursue strategies which appear to be sub-optimal and which represent iterations of these automatic responses. This stringing together of actions initially derived from low level plans seems to inhibit the formation of higher level plans involving more efficient methods.

Unfortunately, the question of what constitutes an "optimal" sequence of edit commands is quite problematic. Embley & Nagy (1982) report that "experts" can often reduce the number of keystrokes used to perform some task by a factor of three compared with relative novices. However, each expert took about an hour to devise such command sequences, and optimization generally involved combinations that even experienced users would consider unusual. The two main issues appear to be: (1) we do not know how to characterize "optimal" from the point of view of user psychology and (2) the effort towards analysing complexity in command sequences might be better directed towards the design of more principled command languages. For example, Anandan, Embley & Nagy (1980) have suggested that file comparison algorithms might prove useful in comparing the command structure of editors and in producing new command sets.

Thus promoting plans which have proved successful in the past is not always the most productive way to facilitate learning. The program would need to be able to rate plans for achieving some goal according to their relative efficiency, possibly by employing some kind of ends–means analysis and then computing the cost of the various methods. On the other hand, the program may volunteer action sequences which strike users as odd and do not fit either their conception of the task or their mental model of the way in which the system works. An advice-giving program would have to be capable of a good deal of introspection in order to be sensitive to such issues; for example, it would need access to a representation of its own theory of representation, as well as a representation of the user's model of itself. This goes beyond reasoning about control, since it involves reasoning about the conventions for coding knowledge, as well as the conventions for applying it. Nevertheless, one feels that an intelligent system ought to be able to reflect upon every aspect of its own function, and one suspects that this ability is crucial for most forms of learning, including plan generalization. The interested reader is referred to Batali's (1983) review of computational systems which exhibit introspective behaviour.

To summarize, the fundamental idea behind the fusion of planning, speech act interpretation and meta-level inference put forward in this paper is that a program should be able to entertain the propositional contents of questions and commands without necessarily either answering the question or obeying the command. For example, a "how" query, such as "how do I delete TEXT.DOC?", where TEXT.DOC does not exist, might be answered with a query, or an assertion of this fact, rather than a plan. Similarly, when evaluating commands like "delete TEXT.DOC!", we would like to project their consequences and check that they do not violate constraints before executing them. This ability to reason about possible states of affairs before trying to make them actual is a prerequisite for any planning system that wishes to incorporate commonsense reasoning in this domain (see, for example, Faletti, 1982).

5. Prospects of the present approach and further problems

Norman (1983) has pointed out that commands typically confuse intentions and actions. Users of computing systems start with intentions, choose actions which are apt to achieve their goals, specify these actions and evaluate the result. Indicating one's intention by simply typing a command can have a number of negative outcomes: (1) the command may not in fact realize one's intention; (2) the command realizes the intention but has an unwanted side effect or (3) the intention was originally ill-formed and not what one really wanted to do. Norman argues that human–computer interfaces should support the intention stage (by providing access to relevant information or history of previous actions), help users in the planning phase, and remind them where they are in the execution phase. Also, user operations should be construed as iterations towards a goal, to which each input is an approximation, rather than a wholly right or wrong step.

We suggest that these desiderata require the interface to exhibit at least some of the planful behaviour outlined in the previous section, with regard to generating and recognizing plans which fit the user's intention. Giving the user information about the pre- and postconditions of achieving some goal is one way of helping him check whether he really wants to proceed with it. It could be that the preconditions are not satisfiable, e.g. not enough space on disk to copy a particular file, or that the postconditions contain unwanted effects, e.g. overwriting an old file with the same name.

An interface driven by the kinds of deduction we have described could possibly stand between the user and the command processor, helping him to refine his intentions and realise his goals. However, it is clear that more than the mere generation of plans or proofs is required to provide the user with the assistance he needs. Activities need to be structured, because the order in which actions occur to the user need not be the order in which commands should be executed in the interests of safety and efficiency.

In the design of an intelligent interface, one might be interested in propositional attitudes other than "how do I do p?" or "do p!". For example, one might wish to indicate "I'm thinking of doing p" or "what if I did p?". Allowing the user to be as tentative or assertive as he chooses in his dealings with the system might be one way of providing graded help to users of varying degrees of experience and expertise.

Consider the following example. Suppose the user is able to indicate "I'm thinking of doing p", e.g. "I'm thinking of deleting TEXT.DOC". The interface could then adopt the role of advisor, in pointing out the conditions and consequences of the proposed course of action, identifying problems as they arise. TEXT.DOC might be an ambiguous file specification, in which case the advisor should try and determine which of the matching specifications is intended, rather than simply signalling an error, and supply information to help the user make up his mind about which file he means, such as date of file creation, size of file, position in file structure, and so on. Underspecification constitutes a violation of the preconditions of deletion, and so further specification concerning physical devices or access paths is simply an iteration towards the goal which could be accommodated along the lines of a "query the user" facility.

Other problems that arise on the way to deletion might include file protection and what to do with backups. It is irritating to type a long file specification and then be told that the file cannot be deleted because it is protected, necessitating a retyping of the specification for both the reprotection and the delete commands. Also, one might want to reinstate the backup as the current file, or delete it as well, or rename it or reprotect it. These options represent part of the postconditions of an operation; decisions which have to be taken as a result of the perturbation which the command would introduce to the system. Other postconditions include the release of a certain amount of storage space, which may well have been the user's original motivation for the deletion. The amount of space released might have a bearing upon subsequent deletions or reinstatements, and should therefore be displayed.

The advantages of this kind of interaction are as follows: (1) there are no "errors", only hypothetical conditions and consequences, at least until the user ratifies his intention and its associated outcomes; (2) the user is allowed to "think around" what he wants to do, without having a particular order imposed on him, e.g. check backups, reprotect, delete, rename, etc. and (3) the execution of irreversible commands can be delayed until all the associated decisions to do with the final outcome have been taken.

Allowing the user to be tentative in his approach is one way of grading help. Another way might be to tailor the level of detail in the specification of plans, by maintaining a model of the user which takes note of which tasks he can typically perform unaided, and then using this information to inhibit plan expansion. If the user has regularly marked blocks successfully, then he does not need to be told how to do this when planning a cut and paste, and the knowledge base could contain a meta-rule which generalizes this insight. Also, the emission of warnings could be made more responsive to individuals by keeping a record of past mistakes. Such responses could be very specific, as in individualized spelling correctors, or more general, e.g. this user has trouble handling files, so file warnings will be given in full. However, it is difficult to know at what point the system should actually interrupt the user's activity, even if one has a user model of some kind [although see Burton & Brown (1982) for guidelines].

Finally, the uniformity and modularity of many rule-based systems might help solve some of the problems associated with keeping help up to date with system development. If the same rules are used to manage the interaction, generate plans, provide explanations of program behaviour, and so on, then augmenting one of these functions by adding a new rule should augment each of the other functions associated with a different procedural interpretation of such rules. Accordingly, the interface could be viewed as a self-documenting system, whereby extensions to the system automatically extend its explanatory power.

6. Conclusions

The proposed application of rule-based systems to the interface between user and utility program leaves many of the practical and theoretical issues raised in the introduction unaddressed. For example, users need "metahelp", i.e. help in using help facilities, and there remains the difficult question of what constitutes a good mental model of a system [see, for example, papers in Gentner & Stevens (1982)]. Also, it is difficult to extend system capabilities without increasing system complexity. Giving users access to support facilities should not result in a proliferation of modes and menus. As O'Shea & Self (1983, p. 182) point out, a system with a "multiplicity of roles, as tutor, interpreter, editor and commentator, can lead to difficulties for the novice who must keep in mind with which of these notional machines he is dealing".

In addition, we have already noted various problems of control associated with the kinds of deduction a rule based system would have to make while monitoring the interaction. Plan recognition and plan generalization pose special difficulties, particularly in more open-ended tasks, where the space of possible ways in which the user could reasonably be expected to proceed is quite large. Also, the inevitability of noise, both at the level of individual actions (e.g. key strokes, mouse clicks, etc.), and at the level of changes in intention (e.g. abandoning goals or rejecting plans) seems to require fairly sophisticated channels of meta communication if the interface is to be at all flexible.

Nevertheless, a rule-based approach to interface design has a number of things to recommend it. (1) The very process of submitting the task to the kind of analysis that the codification of expertise requires is valuable in itself. (2) The use of meta-rules might allow such programs to reason about the way in which they are being used, for the purposes of modelling both the current state of the interaction and (ultimately) the individual user. (3) The uniformity and modularity of rule-like representations lends itself to principled systems extension and revision with the possibility of automating some aspects of help an documentation update.

The work described here was funded under research grants FG60 1123243 and 6331 from the Open University. The authors would like to thank Richard Young and Tim O'Shea for their comments on an earlier draft of this paper. "WordStar" is a trademark of MicroPro International Corporation.

References

ALLEN, J. (1983). Recognizing intentions from natural language utterances. In BRADY, M. & BERWICK, R. C., Eds, *Computational Models of Discourse*, pp. 107–166. Cambridge, Massachusetts: M.I.T. Press.

ANANDAN, P., EMBLEY, D. W. & NAGY, G. (1980). An application of file-comparison algorithms to the study of program editors. *International Journal of Man–Machine Studies*, **13**, 201–211.

AUSTIN, J. L. (1962). *How to do Things with Words*. Oxford: Clarendon Press.

BAR-HILLEL, Y. (1954). Indexical expressions. *Mind*, **63**, 359–379.

BARTSCH, R. (1979). Semantical and pragmatical correctness as basic notions of the theory of meaning. *Journal of Pragmatics*, **3**, 1–43.

BATALI, J. (1983). Computational introspection. *AI Memo No. 701*, Massachusetts Institute of Technology, Artificial Intelligence Laboratory.

BELNAP, N. D. & STEEL, T. B. (1976). *The Logic of Questions and Answers*. Yale: Yale University Press.

BURTON, R. R. & BROWN, J. S. (1982). An investigation of computer coaching for informal learning activities. In SLEEMAN, D. & BROWN, J. S., Eds, *Intelligent Tutoring Systems*, pp. 79–98. London: Academic Press.

CHESTER, D. (1976). Translating mathematical proofs into English. *Artificial Intelligence*, **6**, 261–278.

CLANCEY, W. J. (1981). Methodology for building an intelligent tutoring system. *Memo. HPP-81-18*, Stanford Heuristic Programming Project, Stanford University, California.

CLANCEY, W. J. (1982). Tutoring rules for guiding a case method dialogue. In SLEEMAN, D. & BROWN, J. S., Eds, *Intelligent Tutoring Systems*, pp. 201–225. London: Academic Press.

CLANCEY, W. J. & LETSINGER, R. (1982). NEOMYCIN: Reconfiguring a rule-based expert system for application to teaching. *Report No. STAN-CS-82-908*, Department of Computer Science, Stanford University.

COHEN, P. R. & PERRAULT, C. R. (1979). Elements of a plan-based theory of speech acts. *Cognitive Science*, **3**, 177–212.

COOMBS, M. J. & ALTY, J. L. (1980). Face-to-face guidance of university computer users—II: Characterising advisory interactions. *International Journal of Man–Machine Studies*, **12**, 407–429.

CRESSWELL, M. J. (1973). *Logics and Languages*. London: Methuen.

CULLINGFORD, R. E. & KRUEGER, M. W. (1980). Automated explanations as a component of a computer-aided design system. *IEEE Transactions on Systems, Man & Cybernetics*, **SMC-10**, 343–349.

CULLINGFORD, R. E., KRUEGER, M. W., SELFRIDGE, M. G. & BIENKOWSKI, M. A. (1982). Automated explanations as a component of a computer-aided design system. *IEEE Transactions on Systems, Man & Cybernetics*, **SMC-12**, 168–181.

DAVIS, R. (1983). TEIRESIAS: experiments in communicating with a knowledge-based system. In SIME, M. E. & COOMBS, M. J., Eds, *Designing for Human–Computer Communication*, pp. 87–137. London: Academic Press.

DAVIS, R. & LENAT, D. (1982). *Knowledge-based Systems in Artificial Intelligence.* New York: McGraw-Hill.

EMBLEY, D. W. & NAGY, G. (1982). Can we improve text editing performance? *Proceedings of the Conference on Human Factors in Computer Systems,* Gaithersburg, pp. 152–156. Washington, D.C.: Association for Computing Machinery.

FALETTI, J. (1982). PANDORA—A program for doing commonsense planning in complex situations. In *Proceedings of the National Conference on Artificial Intelligence (AAAI-82),* Carnegie–Mellon University, Pittsburgh, Pennsylvania, pp. 185–188.

FIKES, R. & NILSSON, N. (1971). STRIPS: A new approach to the application of theorem proving to problem solving. *Artificial Intelligence,* **2**, 189–208.

FODERARO, J. K. & SKLOWER, K. L. (1981). *The Franz Lisp Manual.* University of California at Berkeley.

GAINES, B. R. & SHAW, M. L. G. (1983). Dialog engineering. In SIME, M. E. & COOMBS, M. J., Eds, *Designing for Human–Computer Communication,* pp. 23–53. London: Academic Press.

GENESERETH, M. R. (1978). Automated consultation for complex computer systems. *Ph.D. thesis,* Harvard University.

GENESERETH, M. R. (1982). The role of plans in intelligent teaching systems. In SLEEMAN, D. & BROWN, J. S., Eds, *Intelligent Teaching Systems,* pp. 137–155. London: Academic Press.

GENESERETH, M. R. (1983). An overview of meta-level architecture. *Proccedings of National Conference on Artificial Intelligence (AAAI-83),* pp. 119–124.

GENTNER, D. & STEVENS, A., Eds (1982). *Mentals Models.* Hillsdale, New Jersey: Erlbaum.

HAYES-ROTH, B. & HAYES-ROTH, F. (1979). A cognitive model of planning. *Cognitive Science,* **3**, 275–310.

HEWITT, C. (1969). PLANNER: A language for proving theorems in robots. *Proceedings of the International Joint Conference on Artificial Intelligence,* Washington, D.C., pp. 295–301.

JAGODZINSKI, A. P. (1983). A theoretical basis for the representation of on-line computer systems to naive users. *International Journal of Man–Machine Studies,* **18**, 215–252.

KASHER, A. (1977). What is a theory of use? *Journal of Pragmatics,* **1**, 105–120.

KENNEDY, T. C. S. (1974). The design of interactive procedures for man–machine studies. *International Journal of Man–Machine Studies,* **6**, 309–334.

KLEMPERER, K. (1982). A user-friendly interactive user assistance system. In *Proceedings of the ASIS 45th Annual Meeting,* p. 368. Columbus, Ohio: American Society for Information Science.

KOWALSKI, R. (1979). *Logic for Problem Solving.* New York: Elsevier/North–Holland.

KOWALSKI, R. (1982). Logic as a computer language. In CLARK, K. L. & TARNLUND, S.-A., Eds, *Logic Programming,* pp. 3–16. London: Academic Press.

LANG, K. & AULD, R. (1982). The goals and methods of computer users. *International Journal of Man–Machine Studies,* **17**, 375–399.

LENAT, D., DAVIS, R., DOYLE, J., GENESERETH, M., GOLDSTEIN, I. & SCHROBE, H. (1981). Meta-cognition: Reasoning about knowledge. *Memo. HPP-81-21,* Stanford Heuristic Programming Project, Stanford University, California.

LEWIS, B. N. & COOK, J. A. (1969). Towards a theory of telling. *International Journal of Man–Machine Studies,* **1**, 129–176.

LEWIS, D. K. (1972). General semantics. In DAVIDSON, D. & HARMAN, G., Eds, *Semantics of Natural Language.* Dordrecht: Reidel.

LONDON, R. & CLANCEY, W. J. (1982). Plan recognition strategies in student modelling: prediction and description. In *Proceedings of the National Conference on Artificial Intelligence (AAAI-82),* Carnegie–Mellon University, Pittsburgh, Pennsylvania, pp. 335–338.

MARTIN, J. (1973). *Design of Man–Computer Dialogues.* Englewood Cliffs, New Jersey: Prentice-Hall.

MCCARTHY, J. & HAYES, P. J. (1969). Some philosophical problems from the standpoint of artificial intelligence. In MELTZER, B. & MICHIE, D., Eds, *Machine Intelligence 4,* pp. 463–502. New York: Edinburgh University Press.

MILLER, M. L. (1982). A structured planning and debugging environment for elementary programming. In SLEEMAN, D. & BROWN, J. S., Eds, *Intelligent Teaching Systems*, pp. 119–136. London: Academic Press.

MONTAGUE, R. (1972). Pragmatics and intensional logic. In DAVIDSON, D. & HARMAN, G., Eds, *Semantics of Natural Language*. Dordrecht: Reidel.

MOON, D., STALLMAN, R. & WEINRAUB, D. (1983). *Lisp machine manual* (5th Edition). Massachusetts Institute of Technology, Artificial Intelligence Laboratory.

MORRIS, C. W. (1971). *Writings on the General Theory of Signs*. The Hague: Mouton.

NAIMAN, A. (1982). *Introduction to WordStar*. Berkeley: Sybex.

NEIMAN, D. (1982). Graphical animation from knowledge. In *Proceedings of the National Conference on Artificial Intelligence (AAAI-82)*, Carnegie–Mellon University, Pittsburgh, Pennsylvania, pp. 373–376.

NEWELL, A. & SIMON, H. A. (1976). Computer science as empirical inquiry: Symbols and search. *Communications of the Association for Computing Machinery*, **19**, 113–126.

NORMAN, D. (1981). Categorization of action slips. *Psychological Review*, **88**, 1–15.

NORMAN, D. (1983). Tradeoffs in the design of human computer interfaces (unpublished).

O'SHEA, T. & SELF, J. (1983). *Learning and Teaching with Computers*. Brighton: Harvester Press.

RESCHER, N. (1966). *The Logic of Commands*. London: Routledge & Kegan Paul.

SACERDOTI, E. D. (1974). Planning in a hierarchy of abstraction spaces. *Artificial Intelligence*, **5**, 115–135.

SACERDOTI, E. D. (1975). The non-linear nature of plans. *Proceedings of the Fourth International Joint Conference on Artificial Intelligence*, pp. 206–214.

SCHANK, R. C., Ed. (1975). *Conceptual Information Processing*. New York: North-Holland.

SCHANK, R. C. & ABELSON, R. P. (1977). *Scripts, Plans, Goals and Understanding*. Hillsdale, New Jersey: Erlbaum.

SCHMIDT, C. F., SRIDHARAN, N. S. & GOODSON, J. L. (1978). The plan recognition problem. *Artificial Intelligence*, **11**, 45–83.

SEARLE, J. (1969). *Speech Acts: An Essay in the Philosophy of Language*. New York: Cambridge University Press.

SERGOT, M. (1983). A query-the-user facility for logic programming. In DEGANO, P. & SANDEWALL, E., Eds, *Integrated Interactive Computing Systems*, pp. 27–41. Amsterdam: North-Holland.

SHRAGER, J. & FININ, T. (1982). An expert system that volunteers advice. In *Proceedings of the National Conference on Artificial Intelligence (AAAI-82)*, Carnegie–Mellon University, Pittsburgh, Pennsylvania, pp. 339–340.

SIME, M. E. & COOMBS, M. J., Eds (1983). *Designing for Human–Computer Communication*. London: Academic Press.

STALLMAN, R. M. (1979). EMACS: The extensible, customizable, self-documenting display editor. *AI Memo. 519*, M.I.T., Cambridge, Massachusetts.

STEVENS, A., ROBERTS, B. & STEAD, L. (1982). The use of a sophisticated graphics interface in computer assisted instruction (mimeo). Presented at the *National Computer Graphics Association Conference*, Anaheim, California, June. Cambridge, Massachusetts: Bolt Beranek & Newman.

SULLIVAN, M. A. & CHAPANIS, A. (1983). Human factoring a text editor manual. *Behaviour and Information Technology*, **2**, 113–125.

TANAKA, H., CHIBA, S., KIDODE, M., TAMURA, H. & KODERA, T. (1982). Intelligent man–machine interface. In MOTO-OKA, T., Ed., *Fifth Generation Computer Systems*, pp. 147–157. Amsterdam: North-Holland.

THOMAS, J. C. & CARROLL, J. M. (1981). Human factors in communication. *IBM Systems Journal*, **20**, 237–263.

WILENSKY, R. (1982). Talking to UNIX in English. In *Proceeding of the National Conference on Artificial Intelligence (AAAI-82)*, Carnegie–Mellon University, Pittsburgh, Pennsylvania, pp. 103–105.

Expert systems and information retrieval: an experiment in the domain of biographical data management

GIAN PIERO ZARRI

Centre National de la Recherche Scientifique, Laboratoire d'Informatique pour les Sciences de l'Homme, 54, Boulevard Raspail, 75270 Paris Cedex 06, France

The RESEDA project is concerned with the construction of Artificial Intelligence (AI) management systems working on factual databases consisting of biographical data; this data is described using a particular Knowledge Representation language ("meta-language") based on the Artificial Intelligence understanding of a "Case Grammar" approach. The "computing kernel" of the system consists of an inference interpreter. Where it is not possible to find a direct response to the (formal) question posed, RESEDA tries to answer *indirectly* by using a first stage of inference procedures ("transformations"). Moreover, the system is able to establish automatically *new* causal links between the statements represented in the base, on the ground of "hypotheses", of a somewhat general nature, about the class of possible relationships. In this case, the result of the inference operations can thus modify, at least in principle, the original content of the database.

1. Introduction

This article aims to provide a brief description of RESEDA's conceptual tools. RESEDA is an expert system equipped with "deep level" (Davis, 1982; Hart, 1982) reasoning capabilities in the field of complex (structured) biographical data management. In this respect, RESEDA constitutes a novelty in the field of expert systems where, to the author's knowledge, this type of data has never before been considered; for works which, in the far wider field of research in Artificial Intelligence (AI), have approached the elaboration of information which could generally be treated as "biographical" see, for example, Carbonnel (1978), Kolodner (1980), Hafner (1981), and McCarty (1982). In the present state of the system, the biographical information processed by RESEDA concerns a well defined period in time (approximately between 1350 and 1450) and a particular subject area (French history).

A complete prototype of the RESEDA system has been operational for at least a year. This prototype is written in VSAPL and is implemented at the "Centre Inter-Régional de Calcul Electronique" (CIRCE) at Orsay, France.

2. Fundamental concepts of the RESEDA system

RESEDA rightly comes into the category of expert systems. From a general methodological point of view, it utilizes knowledge inherent in a *well-defined subject area* (mediaeval French history) to achieve the concrete objective of simulating an *expert's behaviour* (intelligent information retrieval) within this field. It conforms, therefore, to the well-known paradigm of "knowledge engineering" (Goldstein & Papert, 1977; Feigenbaum, 1977).

DEVELOPMENTS IN EXPERT SYSTEMS
ISBN 0-12-187580-6

This coherence of purpose is also reflected on a more technical level. For example, RESEDA's inference rules can be considered as pertaining to the general framework of the type "situation → action", like the "production rules" used in most expert systems (Michie, 1979). These rules are selected by means of a mechanism which, though simple (see section 5.3), is similar to that used for the "metarules" (Davis, 1977, 1980). The rules are then executed by an "inference interpreter". In realizing this, the usual problems which arise in this area had to be solved; for example, the problem of backtracking (see section 5.2).

On the other hand, RESEDA shows a certain number of original characteristics compared with the "average" expert system. As it belongs to the domain of information retrieval, its database—where the "facts" (biographical information) which the system is able to process are stored—is obviously permanent and extensive. This contrasts with most expert systems where the information to be interpreted is introduced at the time of processing. Moreover, this information is represented in an advanced "metalanguage" (see section 3), which is richer and more expressive than the very inflexible system using "attribute–object–value" triples to which most expert systems are confined (see Van Melle, 1981). RESEDA's originality with respect to data representation is particularly significant because, as will become apparent later, all other kinds of knowledge representation used in the system (e.g. those for the inference rules), are simply generalizations of the kind adopted for the data.

In section 3 of this article the theoretical principles of the metalanguage and the criteria for its practical utilization are described. Sections 4 and 5 are devoted, on the other hand, to the system's inference procedures. In section 6, "Conclusion", I shall give information about some preliminary studies concerning the feasibility of adapting RESEDA's methodology to the processing of biographical data other than historical.

3. The metalanguage

The form of representation selected for the "facts" which constitute the system's database has been chosen taking into account two general constraints.

In RESEDA, the term "biographical data" must be understood in its widest possible sense: being, in fact, any event in the public or private life, physical or intellectual, that it is possible to gather about the people in whom we are interested and their context. The form of representation adopted must thus allow us to represent this richness with a minimum of information loss.

At the same time, the construction of the database is not an end in itself, independent of the general aim of the system: the data representation must thus allow a certain amount of efficiency, from a computing point of view, to be achieved during the retrieval operations performed inside the base.

The search for a compromise between these two contradictory requirements lead us to adopt, for the database, a solution based on the use of a strictly controlled lexicon, utilized in pre-defined syntactic patterns constructed from semantic primitives. The solution adopted is therefore of a "metalanguage" type rather than, for example, of the "unrestricted natural language" type. Our metalanguage essentially follows the particular "Case Grammar" conception as developed by AI researchers from the ideas of Fillmore (1968, 1977), i.e. adapted to the structuring of deep semantic "contents" and not of surface structures (Bruce, 1975; Goldstein & Papert, 1977; Rosner & Somers, 1980; Charniak, 1981; etc.). Extensions to classical Case Grammar—especially concerning the representation of logical links between the elementary units structuring the biographies ("coded episodes" or "planes")—allow the metalanguage to attain a certain degree of flexibility and richness as called for by the first of the requirements mentioned above. It is reasonable here to quote a certain number of intellectual contributions to the progressive development of the metalanguage: Silvio Ceccato's theory of correlators (1961); certain evolved documentary languages (Gardin, 1965); the work of Schank (Schank, 1973, 1975; Schank & Abelson, 1977).

3.1. AN EXAMPLE OF CODING

In order to illustrate certain fundamental characteristics of the metalanguage, an example of a set of biographical information and its coded representation are given—see Fig. 1. This information, which was obtained by analysing the work of L. Douët-d'Arcq, *Choix de pièces inédites relatives au règne de Charles VI*, concerns the biographies of Jeanne de Chaumont, Henry Champaigne and King Charles VI:

> During the wars between Charles VI and the English, in the years following Pentacost 1388, a bourgeois family from Périgueux, that of Jeanne de Chaumont, was completely deprived of its possessions by the English troops stationed at Montagrier. Only after some time did Jeanne de Chaumont resign herself to negotiating her marriage with an English squire named Henry Champaigne, on the condition that funds be sent to her family. Considering this, forgiveness for having an English husband was granted to her by King Charles VI in May 1399.

This passage was broken down into seven "elementary episodes"; the natural language formulation that accompanies them in Fig. 1 simply attempts to help to locate them within the original complete information.

The coded episodes (or "planes") in Fig. 1 are not enough, however, to translate the original information completely: it is, in fact, necessary to add an eighth plane, see Fig. 2, which provides the logical links existing between the data in Fig. 1.

The brackets in Fig. 2 are added here merely for the sake of legibility, to show more clearly the main logical elements of the expression: 7 happened because 2 was caused by 1. The data in parentheses provide the extra details. In reality, in the coding as well as in the elaboration, there is no hierarchy between the "lists" introduced by one of the codes in Fig. 2 (CONFER, CAUSE, FINAL) (for an explanation of these codes, see section 3.2.3).

3.2. CHARACTERISTICS OF THE METALANGUAGE

The planes given in the previous section will now allow us to proceed with a simple discursive description of RESEDA's metalanguage.

3.2.1. Predicative planes

Planes of type 1–2–3–4–5–7 are called "predicative planes"; plane 6 is a "relational plane" and plane 8 in Fig. 2 a "parenthetical plane". The coding of predicative planes is closest to classical Case Grammar; each predicative plane is constructed around one, and only one, predicate—BE-AFFECTED-BY (planes 1, 3, 4, 5), BEHAVE (plane 2), and PRODUCE (plane 7) in Fig. 1—accompanied by indicators of the case, called "correlators" in our terminology. The correlators introduce the "arguments" of the

(1) end+BE-AFFECTED-BY SUBJ   household (SPECIF Jeanne-de-Chaumont):
                                               Périgueux
                OBJ   goods (SPECIF all)
                SOURCE   troops (SPECIF political-entity (SPECIF
                                                 England)) : Montagrier
                EVENT   hundred-year-war
                date1:   between-27-may-1388-and-(may-1393)
                date2:
                bibl.:   Douët-d'Arcql, 154

"Jeanne de Chaumont's family is completely ruined by the English"
(2) BEHAVE SUBJ   Jeanne-de-Chaumont: Périgueux
        OBJ   Henry-Champaigne
        MODAL   negotiation
        date1:   between-(may-1393)-and-(may-1396)
        date2:   between-(may-1396)-and-(may-1398)
        bibl.:   Douët-d'Arcql, 154

"Jeanne de Chaumont negotiates with Henry Champaigne"
(3) *begin+recip+BE-AFFECTED-BY SUBJ   (COORD Jeanne-de-Chaumont
                                                        Henry-Champaigne)
                                OBJ   (COORD Jeanne-de-Chaumont
                                                        Henry-Champaigne)
                              MODAL   marriage
                              date1:
                              date2:
                              bibl.:   Douët-d'Arcql, 154

"Jeanne de Chaumont and Henry Champaigne would be married"
(4) *BE-AFFECTED-BY SUBJ   household (SPECIF Jeanne-de-Chaumont):
                                            Périgueux
              OBJ   money
              date1:
              date2:
              bibl: Douët-d'Arcql, 154

"Jeanne de Chaumont's family would have money"
(5) const+BE-AFFECTED-BY SUBJ   army (SPECIF political-entity
                                            (SPECIF England))
                          OBJ   Henry-Champaigne (SPECIF squire)
                          date1:   between-(may-1396)-and-(may-1398)
                          date2:
                          bibl.:   Douët-d'Arcql, 154

"Henry Champaigne is a squire in the English army"
(6) (COORD Jeanne-de-Chaumont (SPECIF spouse) Henry-Champaigne (SPECIF spouse))
"Henry Champaigne is Jeanne de Chaumont's spouse"
(7) PRODUCE SUBJ   Charles-VI: Paris
           OBJ   remission
           DEST   Jeanne-de-Chaumont: Périgueux
           date1:   may-1399
           date2:
           bibl.:   Douët-d'Arcql, 154

"Jeanne de Chaumont is granted remission"

FIG. 1. Coded representation of the "elementary episodes" corresponding to the natural language text concerning the downfall of Jeanne de Chaumont's family.

(8)   7(CONFER 5 6) [CAUSE 2 (FINAL 3 (FINAL 4)) [CAUSE 1]]

"remission (7) *on the subject of* (CONFER) having a husband (6) belonging to the English army (5) granted (7) *because* (CAUSE) behaviour (2) *having for aim* (FINAL) the wedding (3) *in order to* (FINAL) obtain the resolvency of the family (4) *was caused* (CAUSE) by the ruin of the family (1)"

FIG. 2. Coded representation of the logical links between the "coded episodes" ("planes") in Fig. 1.

predicate. In the structure of RESEDA's metalanguage, the arguments associated with the predicative correlators SUBJ(ect), OBJ(ect), SOURCE and DEST(ination) play a particularly important part and are thus the only ones which can be endowed with a location, for example "Périgueux" and "Montagrier" in plane 1. EVEN(t), "in the context of", and MODAL(ity) are also predicative correlators, whilst SPECIF(ication) and COORD(ination) do not directly introduce arguments of the predicate, but only "expansions" inside the arguments, realized in the system in the form of "lists".

RESEDA's Case System is characterized by an extremely restricted number of predicates—five in all (BE-PRESENT and MOVE do not appear in the example); this is compensated by their association with a group of "modulators", such as "end, begin, recip, const". The modulators can be combined together—as, for example, "begin + recip" in plane 3. The modulators, as their name suggests, are there to specify semantic variations allowed for each predicate, whilst keeping their number to a minimum.

Looking more closely at the different predicative structures constructed around BE-AFFECTED-BY in planes 1, 3 and 5 by the use of three "temporal modulators", "end", "begin" and "const", illustrates the differentiating role played by the modulators associated with a predicate. In the three cases, as when used alone in plane 4, the predicate BE-AFFECTED-BY defines a situation where the SUBject has at its disposal an OBject: inanimate goods or a person. However, in plane 4, this situation is considered—as for BEHAVE in plane 2 and PRODUCE in plane 7—extensively, whilst in planes 1, 3 and 5 it is reduced to a particular privileged point in time: with the modulator "end" (plane 1), the end of the situation; with the modulator "begin" (plane 3) the beginning; and, with the modulator "const" (from the French *constater* = to find, to recognize, plane 5) the moment when the presence of a particular situation is observed. This group of temporal modulators specifies the bearing of date indications carried by predicative planes, an important factor in the logical classification of the data. They are thus recorded in great detail in two date blocks (date1 and date2) which may each contain a date (date1, plane 7 for example) or a fork of dates (date1, plane 1 for example). Their presence is essential: when the date indications are not directly provided by the original documents, they must be "reconstructed" by the historian documentalist, see the dates in parentheses in Fig. 1. Only the first date block is filled when the situation may be represented as a "point" on the time axis (plane 7), or reduced to a point using one of the temporal modulators. The two blocks are filled when a durable situation is considered extensively (plane 2); the two blocks are empty when the situation is "conjectural" (code "*"), that is, presented as an intention (planes 3 and 4, see also section 3.2.3). For a more complete description of the coding of temporal elements in RESEDA, see Zarri *et al.* (1980, pp. 36–70) and Zarri (1983*a*).

Thus, the modulators define a particular connotation of the predicate in a given semantic field: either with respect to a sub-group of modulators, in which case they are mutually exclusive—this is the case of the "temporal" domain above, or the "opinion" domain including the values positive, "for", negative, "against", or neutral, no modulator; or simply by the presence or absence of a modulator, as in the domain of "reciprocity", for example. In this case, the modulator "recip" adds to the normal situation of disposability of the OBJect to the SUBJect in planes with BE-AFFECTED-BY, the meaning of reciprocal disposability between SUBJect and OBJect, as in plane 3.

The different allowed combinations of these three categories of syntactic tools, predicates, correlators and modulators, form the series of "Predicative Schemata" on which depends the coherence and stability of the metalanguage, and which are used in Knowledge Representation throughout the system.

3.2.2. The historical personages and the lexicon

The semantic bearing of the predicative schemata that have just been defined—as well as, elsewhere, the schemata of "relational planes", see plane 6 in Fig. 1, which themselves are represented in a simplified way equivalent to that of an "expansion" in a predicative plane—is only completely defined when the slots associated with the cases are filled by the different elements of the metalanguage allowed for.

In this context, a first category is formed by people's names—here, Jeanne de Chaumont, Henry Champaigne and Charles VI; see, for example, Zarri *et al.* (1980, pp. 28–31) for more about the use of this particular type of information in the structuring of the specifically biographical part of the database and in the retrieval procedures. Toponyms such as "England" belong to the category "free-structure-words"; "Paris", "Montagrier", "Périgueux" are called "INSEE-location" because the numerical code which is assigned to them by the French "Institut National de Statistique" (INSEE) is whenever possible included in their symbolic representation.

All the other terms which appear in the slots of the planes in Fig. 1, "household, goods, all, troops, political-entity, hundred-years-war, negotiation, marriage, money, army, squire, spouse, remission" are elements of RESEDA's lexicon, that is, the part of the database which essentially gives the social and structural background of the period.

The organization of the lexicon is quite complicated, since many types of data representation are used. For example, information on social, economic and bureaucratic structures of the period is given by a set of interconnected general trees; in this case, the data representation is, therefore, only hierarchical. Information about important events, complex situations, ideological trends, etc., cannot be given satisfactorily by this type of coding only, and the representation must be augmented by the use of sets of planes which are included in the biographical part of the database. The definitions in purely hierarchical form and these "definitions-by-plane"—where information to be associated with the terms defined in this way is very subtle and detailed and requires insertion into a very complex logical organization—represent in some ways, inside the structures of RESEDA's lexicon, the two extremes of a scale of complexity. There are, however, other types of lexicon definitions, see below for example "definitions-by-algorithms"; for a detailed description see Zarri *et al.* (1977, pp. 18–21).

To return now to our example, it can be said that the terms "goods, all, political-entity, negotiation, marriage, money, squire, remission" are characterized simply by their insertion in a tree-structure; whilst, on the other hand, "household, troops, hundred-years-war, army" have their meaning fully specified by a particular association with planes recorded in the biographical base proper:

"hundred-years-war", hierarchically defined as pertaining to the sub-tree "conflic-tual-situation", is also defined using planes, established during the creation of the base, which specify its causes and developments;

"troops", which is itself a "specific" of "army", and "household" are inserted in a tree structure which has "group-of-personages" as its root. Because of this, they belong in a way to the category of historical personages. They are thus liable to function as "SUBJect" inside a predicative plane and can be "titulars", like the personages, of a "volume" or "primary index", a file by means of which it is possible to find all the predicative planes in which they appear (Zarri *et al.*, 1980, pp. 24–35; Zarri, 1983*a*).

Finally, "spouse" is classed under the root term "kinship"; the group of lexical items associated with "kinship" is managed by a particular set of algorithms which allow the reconstitution of implicit alliance links by interpreting chains of these items; for this see Ouy *et al.* (1977).

3.2.3. The logic of planes

When inserted in the slots of the planes of the database, all the terms of the lexicon are assigned a validity code—by default equal to "true". In the same way, the planes, once created, are globally assigned a validity code; this is the case, in particular, of the value "conjectural" (code "*") which marks plane 3 and 4 and which appears in natural language as the use of the conditional: "Jeanne de Chaumont and Henry Champaigne *would* be married", "Jeanne de Chaumont's family *would* have money". A plane thus marked no longer represents a "fact" but a "possibility". The case shown in the example concerns the problem of "belief", well known in AI contexts (see, amongst others, Abelson, 1973, 1979; Bruce & Newman, 1978; Cohen & Perrault, 1979). This can be easily understood: as "facts", Jeanne de Chaumont's marriage and Henry Champaigne's funding are not at all "mitigating circumstances" calling for clemency from Charles VI; on the contrary, the reasoning used in the Act of Remission rests on the "intentions" which determined the in itself condemnable behaviour of Jeanne de Chaumont. Thus the guarantees we have about the concrete reality of the matrimonial links which exist between Jeanne and Henry—certified in the original document and translated by plane 6—and, because of this, the likelihood that the money was paid, must not interfere with the very different reality of a previous moment in history when the project was formed.

The apparently simple technique for dealing with these very complex problems becomes fully significant in the context of the system of logical links which associate planes in the biographical base. This coding is taken care of in RESEDA by a system of labelled pointers ("correlator-pointers"), syntactic tools closely linked to the expansion correlators of the predicative organization.

Those which appear in Fig. 2, "CONFER, CAUSE, FINALITY", are all related to the causal field, with different values which are defined by their insertion in RESEDA's own causality taxonomy rather than by reference to natural language terms; it is well known that the representation of causality is one of the fundamental problems facing researchers dealing with Knowledge Representation in an AI context (see, for example, Wilks, 1975, 1977).

Intuitively, this taxonomy is linked to a double dichotomy in the causal field: (a) cause–consequence: (b) weak–strong. But this distribution is, in fact, based on more objective criteria, shown in Fig. 3. From this point of view, the first criterion of the dichotomy is recognizing if the explicative argument—plane (b), introduced by one of the correlator-pointers of the field—is previous to (=CAUSE or CONFER) or subsequent to (=FINAL or MOTIV) the argument that it explains—plane (a)—to which the causal expansion is attached. The second criterion of the dichotomy is based on the traditional distinction between "necessary" and "sufficient"; the coexistence of the two elements characterizes a strong causality (=CAUSE or FINAL) whilst the existence of only one of the two implies weak causality (=CONFER or MOTIV).

What I have exposed about the coding of causal links should clarify the representation of conjecture in RESEDA.

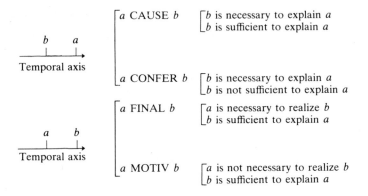

FIG. 3. The "taxonomy of causality" proper to the RESEDA system.

Conjectural planes, marked with an "asterisk" code, are never used to record the historian's conjectures; these are taken into account, of course, in the database, but in an explicit form; for example, a reasoning of this type: "If he had written the letter before Christmas, he would not have been able to mention a present received then, and yet he does mention it" would simply give rise, in the temporal information associated with the plane describing the writing of the letter, to the indication "*terminus-a-quo*-25-december". Here, on the contrary, we have to record the *intentions, hopes* and *conjectures* of the historical personages themselves. A conjectural plane can thus never exist in a totally independent manner, and must necessarily be introduced by a "true" plane describing which personage originated it. For example, it can appear inside a "FINAL" list associated with a "sure" plane, see in the example in Fig. 2 the reciprocal positions of planes "2" (true) and "3, 4" (conjectural); a second way of introducing them, which does not appear in the example, is to consider them as "object" in a construction known as "completive", see Zarri *et al.* (1977, 1980, pp. 10–11). The classical schema for the translation of "intention" in RESEDA is as in Fig. 4 (see again Figs 1 and 2).

More details on the above can be found in Zarri *et al.* (1980, pp. 7–16).

> (a) BEHAVE   SUBJ   personage
> (b) *PREDICATE with any syntax
> (c) a (FINAL b)

FIG. 4. The general schema for the translation of "intention" in RESEDA.

4. RESEDA's inference rules

What I have explained in the previous paragraphs could be seen as pertaining to a totally static, uninteresting and old-fashioned point of view on the problem of Knowledge Representation—see the solutions adopted in the 1960s—if the information of a clearly dynamic type describing the inference rules stored in RESEDA's "Base of Rules" was not formulated using a generalization of the Knowledge Representation language employed for the biographical data itself. This generalization is achieved mainly through the introduction of variables, which allows us to write inference rules applicable to a large class of events. The "restrictions" associated with these variables specify the group of values which can be associated with each of them, and thus define the field where that inference rule can apply (Zarri, 1981).

4.1. THE "SEARCH MODEL" CONCEPT

When the system is considered from the point of view of its utilization, the fundamental concept which must be introduced is that of the "search model". A search model gives the essential elements, expressed in terms of the RESEDA metalanguage, of a coded episode which it is necessary to search for in the database. A search model may originate from outside the system, if it is the direct translation of a query posed by the user. On the other hand, it may be automatically generated by the system, as will be clarified later, during the execution of an inference procedure.

Suppose, for example, that we have introduced into the biographical database the plane in Fig. 5, which is the representation of "Robert de Bonnay was named *bailli*

> (9) begin + soc + BE-AFFECTED-BY   SUBJ   Robert-de-Bonnay
>   OBJ   bailli: Mâcon
>   SOURCE   king's-council: Paris
>   date1:   27-september-1413
>   date2:
>   bibl.:   Demurgerl, 234

FIG. 5. Coded representation of Robert de Bonnay's nomination as *bailli* of Mâcon.

of Mâcon on 27th September 1413 by the King's Council" (biographical authority: Demurger). The *bailli* was a high level civil servant who dispensed justice, administered finances, etc., for a particular area, the *bailliage*, in the name of a king or lord. Now suppose that the user questions RESEDA on the subject of the progression of Robert de Bonnay's career, asking, for example, to keep to a very simple question, "Did Robert de Bonnay exercise the power of a *bailli* during the first quarter of the 15th century". In this case, the user himself creates the search model given in Fig. 6, with the aid of a prompting program. The only notable difference between this formalism and that required for representation of the episodes in the database is that of the presence of a "search interval", "bound1–bound2", which is used to define the temporal

BE-AFFECTED-BY SUBJ   Robert-de-Bonnay
              OBJ   bailli
              bound1:   1400
              bound2:   1425

FIG. 6. A simple "search model" enabling the retrieval of the "plane" in Fig. 5.

limits of the search. Therefore, the search interval has the function of limiting the number of planes to be examined, and has only an indirect relationship with the temporal information of the "date" type which is associated to each of the episodes recorded in the base.

I do not intend, here, to go into the details of the procedure adopted to test the match of a search model with data in the base; see section 5.4.1, below and, for more details, Zarri *et al.* (1979, 1980) and Zarri (1983*a*). It is, for example, obvious that the model in Fig. 6 may be directly matched with the plane in Fig. 5. This, of course, is the exception rather than the rule.

4.2. TRANSFORMATIONS

If we reach a dead end when trying to match a search model with data in the base, a first class of inference rules may be applied; these are the "transformations".

To keep to an extremely simple example, consider the transformation of Fig. 7, allowing us to change a search model formulated in terms of "end + BE-PRESENT" into a new one in terms of "MOVE", which can be submitted, in turn, to the usual match procedures. This formal rule translates the commonsense rule "If someone goes from one place to another, he has certainly left his starting point": the justification of the use of substitution in Fig. 7 lies in the fact that any information about some personage x having moved from k to l is at the same time a response to any query about the possibility of his no longer being at place k. Note that, in the terms of RESEDA's metalanguage, the movements of a personage are always expressed in the form of a subject x which moves itself as an object.

(t1) end + B-PRESENT SUBJ $x:k \rightarrow$ MOVE SUBJ $x:k$
                                      OBJ $x:l$

$x = \langle \text{personage} \rangle$
$k, l = \langle \text{location} \rangle$
$k \neq l$

FIG. 7. A simple inference rule of the "transformation" type.

On a formal level, it is worthwhile noting that the "variables" which appear in the original search model (x and k in Fig. 7) must appear in the transformed model and/or in the "restrictions" associated with the new variables (l in Fig. 7) introduced at the level of this transformed model, see rule t1. This ensures the logical coherence between the two parts of the transformation; the model on the right-hand side must indeed "imply" the one on the left. The values which replace the variables in the retrieved plane (or planes) using the transformed model must obviously respect the restrictions associated with all the variables which appear in the transformation.

A second example of transformation is that given in Fig. 8; the commonsense rule underlying this formalism is: "If a person x has a university degree w, then this person

(t2) PRODUCE SUBJ $y \rightarrow$ BE-AFFECTED-BY SUBJ x
OBJ v OBJ w
DEST x

$x = \langle\text{personage}\rangle$
$y = \langle\text{personage}\rangle|\langle\text{personages}\rangle$
$x \neq y$
$v = \langle\text{university-course}\rangle$
$w = \langle\text{degree-obtained}\rangle$
$w = \text{f}(v)$

FIG. 8. A more complex "transformation" rule.

has followed some course v" (one or several persons y have "produced" the course v with the intention of x). In the transformation in Fig. 8, the variable v of the original search model appears in the terminal model at the level of the restrictions associated with the new variable w; "$w = \text{f}(v)$" is an abbreviated way of expressing that there must be coherence between the diploma obtained and the courses followed.

There is one last remark to be made about transformations, which concerns the two in Fig. 9, written for simplicity's sake in natural language. In this case, the rule about the repercussion of the variables on the left on the right-hand side takes the form of a "condition" which must be met before the transformation can be allowed; the substitution of the model on the right-hand side for the one on the left forces us to check for the existence of episodes within the base which are able to guarantee the appropriate context. Transformations of this kind are called "conditional transformations".

(t3) try to verify the attachment of x to the important person $z \rightarrow$ try to verify the attachment of y to the important person z

on the condition that: verification of a fairly close relationship, for example kinship, between x and y may be found

(t4) try to verify that x belongs to party $v \rightarrow$ try to verify that x fights against party w

on the condition that: it is possible to show that, during the period in question, v and w (e.g. Armagnacs and Bourguignons) were openly opposed

FIG. 9. Examples of "conditional transformations".

I shall conclude by pointing out that the use of concepts comparable with RESEDA's "semantic transformations" is quite common when using AI techniques to exploit a factual database, see for example "elaborations" in Kolodner (1980), "expansions" in McCarty & Sridharan (1980), "extensions" in Hafner (1981), etc.

4.3. HYPOTHESES

Even taking into consideration this first category of inference rules, the behaviour of the system such as it has been described up to now is rather classic in type. There is, however, a second, more original way of searching in RESEDA: it is possible to search for the hidden "causes", in the widest sense of the word, of an attested fact in the base. For example, if the user, in submitting the query in Fig. 6, obtained in reply the plane in Fig. 5, and if we assume that the "reasons" for the nomination are not explicitly recorded in the database, we will now be able to ask the system to produce a *plausible* explanation of this fact automatically by using a second category of inference rules, the "hypotheses".

In order to give an initial idea, on an intuitive level, of the functioning of the hypotheses, Fig. 10 shows the formulation in natural language of four characteristic hypotheses of the RESEDA system. The first part of each of these rules corresponds to a particular class of confirmed facts (planes) for which one asks the "causes". For example, the plane in Fig. 5 is clearly an exemplification of the first part of the fourth hypothesis in Fig. 10. In RESEDA's terminology, the formal drafting of this first part is called a "premise". The second part (the "condition") gives instructions for searching the database for information which would be able to justify the fact which has been matched with the premise. That is, if planes matching the particular search models which can be obtained from the "condition" part of the hypothesis can be found in the database, it is considered that the facts represented by these planes *could* constitute the justification for the plane-premise and are then returned as the response to the user's query. When trying to match the models obtained from the condition the system can, of course, use inferences of the type "transformation" (see also section 5.1).

Let us now look in some detail at the hypothesis h4 of Fig. 10. A whole family of inference rules expressed in RESEDA metalanguage corresponds in reality to the natural language formulation given in h4; one of these realizations is shown in Fig. 11. A description of the procedure followed to isolate the elements of these families can be found in Zarri (1981) [see also Zarri (1979) for the general methodology for constructing hypotheses].

The meaning, in clear, of the formalism in Fig. 11 is as follows (see also h4 in Fig. 10): to explain what brought the administration n to give post m to x, the hypothesis suggests we check in the system's database for the following two facts, which must be verified simultaneously (operator " $\wedge$ ", "and"):

(h1) . . . one might cease to act on behalf of some other person

BECAUSE

one has abused that person's confidence (e.g. by misrepresenting his view to a third party)

(h2) . . . one might leave something (in one's will) to a (religious) community

BECAUSE

one had some special connection with this community

(h3) . . . one might take a particular attitude in an argument

BECAUSE

one has close links with one of the parties in a conflicting situation

(h4) . . . one might be chosen for a (official) post

BECAUSE

one is attached to a very important personage who has just taken power

FIG. 10. Formulation in natural language of four characteristic "hypothesis" rules.

premise: α
(α) begin + soc + BE-AFFECTED-BY SUBJ x
                                           OBJ m
                                           SOURCE n
                                           date1: $d1$
                                           date2:

restrictions on the variables of the premise schemata:

  $x = \langle personage \rangle$
  $m = \langle monarchic\text{-}post \rangle | \langle seigniorial\text{-}post \rangle$
  $n = king's\ council\ |\ lord's\text{-}council$
if $m = \langle monarchic\text{-}post \rangle$ then $n = king's\text{-}council$
if $m = \langle seigniorial\text{-}post \rangle$ then $n = lord's\text{-}council$

condition: A ∧ B

(A) BE-AFFECTED-BY SUBJ   p(SPECIF y)
                              OBJ x
                              bound1:   b1
                              bound2:   b2

(B) begin + lid + BE-AFFECTED-BY SUBJ n
                                           OBJ   y
                                         bound1:   b3
                                         bound2:   b4

restrictions on the variables of the condition schemata:
$b1 < d1 < b2$
$b3 < d1 = b4$
$y = \langle personage \rangle$
$x \neq y$
$p = \langle seigniorial\text{-}organization \rangle$

FIG. 11. Formal representation of one of the "hypotheses" which can explain the nomination of a high level civil servant.

(A) x was employed by an important person y [the seigniorial administration p specific (SPECIF) to y was "augmented" by x] during a period which includes the time when x was nominated to post n;

(B) at a date that coincides with, or is previous to, the date of nomination $d1$, the administration n comes under the leadership of y [n starts to have y for chief (lid = leader)].

Figure 12 gives the planes obtained by means of this hypothesis in the case of the query about Robert de Bonnay's nomination. The formulation in the metalanguage of this event of courses matches (see Fig. 5) the premise in Fig. 11; these planes can thus provide a plausible explanation for the nomination.

5. Some information about the computer structures within the RESEDA system

The RESEDA system has two modes of operation: the "data acquisition" mode and the "query" mode.

The former is used to load the system, whether we are concerned with the biographical data, the terms of the lexicon, or the inference procedures (transformations and

(2) BE-AFFECTED-BY SUBJ   prince's-court (SPECIF Charles-d'Orléans):Blois
                     OBJ   Robert-de-Bonnay (SPECIF chamberlain)
                     date1:   8-april-1409
                     date2:   (1415)
                     bibl.:   Demurgerl, 234

"Robert de Bonnay held the post of chamberlain to the duc d'Orleans (the court of Charles d'Orléans, whose residence was at Blois, was 'augmented' by Robert de Bonnay) from 8 april 1409 until 1415 (date not confirmed by documents, but reconstituted by the historian)"

(3) begin + lid + BE-AFFECTED-BY SUBJ   king's-council: Paris
                                   OBJ   (COORD Louis-d'Anjou
                                             Charles-d'Orléans
                                             Jean-de-Bourbon
                                             Dauphin-Louis Jean-de-Berry
                                             Bernard-d'Armagnac): Paris
                                   date1:   1st-september-1413
                                   date2:
                                   bibl.:   consensus

"On the 1st september 1413, the leaders of the faction favourable to the Duc d'Orléans (the future 'Armagnac party') took control ('lid' = leader) of the administration of the state. This information is provided by the 'consensus' of historians, who are specialists in this period"

FIG. 12. "Planes" provided by the "hypothesis" in Fig. 11 as a reply to a question concerning the causes of Robert de Bonnay's nomination (Fig. 5).

hypotheses). This mode is strictly reserved for RESEDA's "system" team.

The "query" mode, on the other hand, is the normal mode for the user to whom the system is available, and for whom it was conceived. The questions of the "standard" user could correspond to the simple consultation of information stored in the database, or to the search (more or less aided) of implicit information by the use of inference procedures.

The query mode is the only one with which I shall be concerned here.

5.1. THE "MACHINES" ASSOCIATED WITH THE DYNAMIC BEHAVIOUR OF THE SYSTEM

The modules which define the behaviour of RESEDA as a "query system" are grouped into four "machines" described very briefly below; for more details see Zarri *et al.* (1981).

A first machine is the "model machine"; I am referring here to the group of modules which enable a search model to be constructed directly from a user question (see section 4.1) since the automatic generation of models within the inference rules (see sections 4.2 and 4.3) belongs to the "transformation" and "hypothesis" machines. These two last machines define the behaviour of the system's "inference interpreter".

The model machine has two main modules:

an editor module enabling the user to construct a search model: its data is provided by the user at the terminal and the result is a search model in internal format;

a module managing the user model file which ensures at least the following tasks:

writing a model provided as a parameter (result: the address of that model in the file);

reading a model given by its address (result: the corresponding model).

If need be, this module can also be used by other machines in the system.

The "match machine" is made up of three main modules; the "plane selector", the "parser" and the "variable assigner". All three have a search model as a starting point.

The plane selector has no other input, and produces a first list of addresses, using only the "index level" (see Zarri *et al.*, 1980, 1981; Zarri, 1983*a*) of the biographical database. The corresponding planes are known to have the same predicate and personages as the search model, and to agree with its search interval.

The parser receives as input the search model and a plane (the address of which is in the list produced by the plane selector). It compares the two to decide whether the plane matches the model or not.

Finally, the variable assigner was developed in relation to the inference procedures: the model is likely to come from a hypothesis or transformation schema, which contains variables. Its role is the same as that of the parser, and additionally to give all possible assignments (i.e. plane "values", see section 5.2) for each variable needed by the originating schema (the position of these variables in the search model is given as additional input).

The "hypothesis machine" consists of two main modules:

a "selection module" which, from a predicative plane P existing in the base, provides a list of addresses of hypotheses liable to explain P and

an "execution module" which, given a predicative plane P and the address of an hypothesis H, displays all the planes offered by H as an explanation of P.

The execution module in turn consists of three sub-modules:

a premise schema is processed by a sub-module EXECPREM;

a condition schema is processed by a sub-module EXECCOND and

EXECPREM and EXECCOND ensure the "forward traversing" in the "choice tree" (see section 5.2, below); the backtracking is ensured, on the other hand, by the sole sub-module REEXEC.

In the same way, the "transformation machine" consists of two main modules:

a "selection module" which, from a search model M, provides a list of addresses of transformations liable to operate on M and

an "execution module" which, from a search model M and a transformation T, provides the model transformed from M by T.

One will also notice that the transformation machine can be called from within the hypothesis machine, that is to say that the search models generated inside a hypothesis from condition schemata can, in the case of a failure, be "transformed" into "semantically" equivalent models.

5.2. MAIN CHARACTERISTICS OF THE EXECUTION MODULE OF THE HYPOTHESIS MACHINE

RESEDA's inference rules, as was stated in section 4, present in their formulation a number of original characteristics with respect to rules used generally in expert systems. Their methods of execution are, on the other hand, completely classical. The search for solutions within the database by means of an inference rule of the "hypothesis" type amounts to, for example, the exploration of a "choice tree". The branches of this tree are defined by the different groups of "values" which can be successfully associated with the variables met successively on the route within the hypothesis, from the premise to the last condition schema. The values in question are data (dates, locations, modulators, personages, terms of the lexicon) belonging to the planes found in the database by means of the search models drawn from the premise or condition schemata. The forward traversing of the tree (EXECPREM and EXECCOND) is of the depth-first type; the backtracking (REEXEC) is systematic however, at least in the existing prototype, so that we have an entire panorama of the successes and failures associated with the different branches of the tree. For the advantages and disadvantages of this manner of proceedings see, for example, Charniak, Riesbeck & McDermott (1980, pp. 140–161).

A number of additional computing problems are connected with the semantic richness of RESEDA's inference rules. For example, to verify that a certain candidate value satisfies the restrictions associated with a variable requires that one has firstly solved the possible "conditional restrictions" of the variable. The "conditional restrictions" are used to specify the relationships which must exist between the groups of values which can be associated with two or more variables. Thus, in the example of Fig. 11 (see the "restrictions on the variables of the premise schemata"), if the position which changes occupant depends on the royal administration, the source of the nomination cannot be a seigniorial council.

5.3. THE SELECTION MODULES OF THE HYPOTHESIS AND TRANSFORMATION MACHINE

I will end this section by saying a few words about the mechanism for the selection of RESEDA's inference rules. This mechanism, as already stated in section 2, is connected, in some ways, with the techniques used in Artificial Intelligence to implement the "metarules". This mechanism has two levels.

First, we use the fact that all the hypotheses (and transformations) are divided into five classes, according to the five predicates accepted by RESEDA's metalanguage of which one must obligatorily appear in the formulation of the premise of each hypothesis (or in the formulation of the left-hand side of each transformation). Given that a hypothesis (or a transformation) can only apply to a plane (a search model) if its premise (its left-hand side) can fully match with the formulation of the plane (the model), the predicate in the latter is sufficient to define the appropriate class of hypotheses (or transformations). Thus, the presence of the predicate BE-AFFECTED-BY in the coded episode concerning the appointment of Robert de Bonnay (Fig. 5) will enable—in the case of an attempt to explain this fact by hypothesis—to eliminate all the hypotheses which do not belong to the class BE-AFFECTED-BY.

The second level of selection is the most interesting. Within each class defined by a predicate, the hypotheses are distributed into subclasses constructed according to a similar "general semantic content" of the premise (from now on, I shall restrict myself to the selection of the hypotheses as, in the context of this section, the expression "left-hand side of a transformation" can always be substituted by the expression "premise of a hypothesis"). Thus, the hypothesis "appointment to a civil service post by an official body" in Fig. 11 will belong to the sub-class of hypotheses which relate, in a general way, to the explanation of the fact that a "subject", in the framework of his professional activity, modulator "soc(ial)", is assigned an "object" provided by a second person (or body).

The global semantic content is given by the elements which provide the basic structure of the coded formulation of the premise, modulators and correlators, discarding the data formed by the terms of the lexicon, or the variables carrying restrictions which can replace them. Thus, the sub-class of the class just mentioned, BE-AFFECTED-BY, will be defined by the following "skeleton": "begin + soc + (BE-AFFECTED-BY) SUBJ OBJ SOURCE".

To conclude, each sub-class of hypotheses is accessible once the formal schema (skeleton) which defines it and which allows direct access to the part of the hypotheses file reserved for this sub-class has been recognized in the plane being processed.

In practice, two tables are associated with each predicate, the "hypothesis schemata table" and the "transformation schemata table". In these tables are listed all the skeletons, in the sense defined in the preceding paragraphs, corresponding to the different types of planes (search models) which can be processed, by hypothesis or by transformation, by means of the inference rules associated with the predicate (see also Zarri *et al.*, 1977, annex 7). Hence the tables define the first level of selection; the skeletons in the tables define the second, since they identify the sub-classes of inference rules (sometimes consisting of just one rule) which refer to a similar semantic category.

It must be noted that, in each table, the skeletons are mutually exclusive, since the identification of a sub-group of the file which corresponds to a well-defined sub-class of hypothesis obviously must not be ambiguous. This means that the tables are partially ordered: if, in a table, there are one or more groups of skeletons characterized by a partial coincidence of their constituent elements, modulators and correlators, the skeletons must be ordered by decreasing complexity within each group; for more details about this, see Zarri *et al.* (1980, pp. 71–75).

6. Conclusion

In this paper, I have provided a brief description of RESEDA's conceptual tools: the RESEDA project is concerned with the construction of AI management systems working on factual databases consisting of biographical data. This data is described using a particular Knowledge Representation language ("metalanguage") based on the Artificial Intelligence understanding of a "Case Grammar" approach. The "computer kernel" of the system consists of an inference interpreter. Where it is not possible to find a direct response to the (formal) question posed, RESEDA tries to answer *indirectly* by using a first stage of inference procedures ("transformations"). Moreover, the system is able to establish automatically *new* causal links between the statements represented in the base, on the ground of "hypotheses", of a somewhat general nature, about the class of possible relationships. In this case, the result of the inference operations can thus modify, at least in principle, the original contents of the database.

The RESEDA project will be developed in two different directions.

In a first approach, which could be seen as "pragmatic", we shall attempt to test the adaptability of the prototype, as it exists today, to other types of biographical data. In this context, an application of RESEDA in the field of medicine has already been defined; this concerns the study of medical files on cancerology, trying to complement the information usually found in such files with personal, cultural, socioeconomical, family or environmental factors which are often responsible for the variability and imprecision of diagnostics. Other possible applications of RESEDA's methodology concern the legal and military domains.

A second, more theoretical direction for the development of RESEDA, will be a generalization of the computational solutions adopted for the system prototype. Studies in this direction are centred round four main themes:

to develop and generalize the system for Knowledge Representation proper to RESEDA, including in this theme the widening of the semantic framework, which defines the hypothesis inference rules, allowing for relationships other than causal ones;

to introduce into the existing inference interpreter the possibility of an "intelligent" execution of the rules, allowing, for example, the execution of a rule doomed to fail to be stopped so as to proceed with a more appropriate one, thus avoiding over-rigid and expensive backtracking;

to automate, as far as possible, the very complex strategy (applied manually at the moment) necessary to isolate and formalize the system's inference rules (Zarri, 1981), in order to achieve some kind of automatic knowledge acquisition and

to enable the database to be constructed at least partly automatically, by developing "machine translation" techniques to pass from an initial formulation in natural language of information to be introduced into the system, to its representation in the terms of RESEDA's metalanguage—preliminary results have already been obtained in this area (see Léon *et al.*, 1982; Zarri, 1983*b*).

The construction of the RESEDA prototype has been financed by grants from the "Délégation Générale à la Recherche Scientifique et Technique" (RESEDA/0, CNRS–DGRST contract n° 75·7·0456), the "Institut de Recherche d'Informatique et d'Automatique" (RESEDA/1, CNRS–IRIA contract n° 78·206), and the "Centre National de la Recherche Scientifique" within the framework of the "Action Thématique Programmée Intelligence Artificielle" (ATP n° 955045).

References

ABELSON, R. P. (1973). The structure of belief systems. In SCHANK, R. C. & COLBY, K. M., Eds, *Computer Models of Thought and Language*. San Francisco: Freeman.

ABELSON, R. P. (1979). Differences between belief and knowledge systems. *Cognitive Science*, **3**, 355–366.

BRUCE, B. (1975). Case systems for natural language. *Artificial Intelligence*, **6**, 327–360.

BRUCE, B. & NEWMAN, D. (1978). Interacting plans. *Cognitive Science*, **2**, 195–233.

CARBONELL, J. G. (1978). POLITICS: Automated ideological reasoning. *Cognitive Science*, **2**, 27–51.

CECCATO, S. (1961). *Linguistic Analysis and Programming for Mechanical Translation*. New York: Gordon and Breach.

CHARNIAK, E. (1981). The case-slot identity theory. *Cognitive Science*, **5**, 285–292.

CHARNIAK, E., RIESBECK, C. K. & MCDERMOTT, D. (1980). *Artificial Intelligence Programming*. Hillsdale, New Jersey: Lawrence Erlbaum Associates.

COHEN, P. R. & PERRAULT, C. R. (1979). Elements of a plan-based theory of speech acts. *Cognitive Science*, **3**, 177–212.

DAVIS, R. (1977). Generalized procedure calling and content-directed invocation. In *Proceedings of the Symposium on Artificial Intelligence and Programming Languages—Special Issue of the ACM Sigart Newsletter*, (64), 45–54.

DAVIS, R. (1980). Meta-rules: reasoning about control. *Artificial Intelligence*, **15**, 179–222.

DAVIS, R. (1982). Expert systems: Where are we? And where do we go from here? *The AI Magazine*, **3**(2), 3–22.

FEIGENBAUM, E. A. (1977). The art of artificial intelligence: Themes and case studies of knowledge engineering. In *Proceedings of the 5th International Joint Conference on Artificial Intelligence—IJCAI/5,* Cambridge, 1977. Pittsburgh, Pennsylvania: Carnegie–Mellon University Press.

FILLMORE, C. J. (1968). The case for case. In BACH, E. & HARMS, R. T., Eds, *Universals in Linguistic Theory.* New York: Holt, Rinehart and Winston.

FILLMORE, C. J. (1977). The case for case reopened. In COLE, P. & SADDOCK, J. M., Eds, *Syntax and Semantics*, vol. 8. New York: Academic Press.

GARDIN, J. C. (1965). *SYNTOL* (Rutgers Series on Systems for the Intellectual Organization of Information). New Brunswick: Rutgers University Press.

GOLDSTEIN, I. & PAPERT, S. (1977). Artificial intelligence, language, and the study of knowledge. *Cognitive Science,* **1**, 84–123.

HAFNER, C. D. (1981). *An Information Retrieval System Based on a Computer Model of Legal Knowledge.* Ann Arbor, Michigan: UMI Research Press.

HART, P. E. (1982). Directions for AI in the eighties. *ACM Sigart Newsletter*, (79), 11–16.

KOLODNER, J. L. (1980). Retrieval and organization strategies in conceptual memory: a computer model. *Research Report n° 187,* Yale University Computer Science Department, New Haven.

LÉON, J., MEMMI, D., ORNATO, M., POMIAN, J. & ZARRI, G. P. (1982). Conversion of a French surface expression into its semantic representation according to the RESEDA metalanguage. In HORECKY, J., Ed., *COLING 82—Proceedings of the Ninth International Conference on Computational Linguistics.* Amsterdam: North–Holland.

MCCARTY, L. T. (1982). A computational theory of Eisner V. Macomber. In CIAMPI, C., Ed., *Artificial Intelligence and Legal Information Systems,* vol. 1. Amsterdam: North–Holland.

MCCARTY, L. T. & SRIDHARAN, N. S. (1980). The representation of conceptual structures in TAXMAN II, Part One: Logical templates. *Report LRP-TR-4,* Rutgers University Laboratory for Computer Science, New Brunswick.

MICHIE, D. (1979). *Expert Systems in the Micro Electronic Age.* Edinburgh: University Press.

OUY, G., ZARRI, G. P., ORNATO, M., ZWIEBEL, A., ZARRI-BALDI, L. & BOZZOLO, C. (1977). Project RESEDA/0: Rapport sur les Recherches effectuées du 1er octobre 1976 au 1er avril 1977. *Rapport CNRS/ERHF/1977/DGRST-3,* Equipe Recherche Humanisme Français, Paris.

ROSNER, M. & SOMERS, H. L. (1980). Case in linguistics and cognitive science. *Working Paper n° 40,* Fondazione Dalle Molle, Genève.

SCHANK, R. C. (1973). Identification of conceptualizations underlying natural language. In SCHANK, R. C. & COLBY, K. M., Eds, *Computer Models of Thought and Language.* San Francisco: Freeman.

SCHANK, R. C., Ed. (1975). *Conceptual Information Processing.* Amsterdam: North–Holland.

SCHANK, R. C. & ABELSON, R. (1977). *Scripts, Plans, Goals and Understanding.* Hillsdale, New Jersey: Lawrence Erlbaum Associates.

VAN MELLE, W. J. (1981). *System Aids in Constructing Consultation Programs.* Ann Arbor, Michigan: UMI Research Press.

WILKS, Y. (1975). Seven theses on artificial intelligence and natural languages. *Working Paper n°17,* Fondazione Dalle Molle, Castagnola, Lugano.

WILKS, Y. (1977). What sort of taxonomy of causation do we need for language understanding? *Cognitive Science,* **1**, 235–264.

ZARRI, G. P. (1979). What can artificial intelligence offer to computational linguistics? The experience of the RESEDA project. In AGER, D. E., KNOWLES, F. E. & SMITH, J., Eds, *Advances in Computer-aided Literary and Linguistic Research.* Birmingham: University of Aston.

ZARRI, G. P. (1981). Building the inference component of an historical information retrieval system. In *Proceedings of the 7th International Joint Conference on Artificial Intelligence—IJCAI/7,* Vancouver, 1981. Menlo Park: The American Association for Artificial Intelligence.

ZARRI, G. P. (1983*a*). An outline of the representation and use of temporal data in the RESEDA system. *Information Technology: Research and Development*, **2**, 89–108.

ZARRI, G. P. (1983*b*). Automatic representation of the semantic relationships corresponding to a French surface expression. In *Proceedings of the Conference on Applied Natural Language Processing*, Santa Monica, 1983. Menlo Park: Association for Computational Linguistics.

ZARRI, G. P., ORNATO, M., KING, M., ZWIEBEL, A. & ZARRI-BALDI, L. (1977). *Projet RESEDA/0: Rapport Final*. Paris: Equipe Recherche Humanisme Français.

ZARRI, G. P., ORNATO, M., LEE, G., LELOUCHE, R., MEISSONNIER, V. & INSOLE, A. (1979). Projet RESEDA/1: Rapport sur les recherches effectuées du 1er mai 1979 au 1er octobre 1979. *Rapport LISH/150*, Laboratoire d'Informatique pour les Sciences de l'Homme, Paris.

ZARRI, G. P., ORNATO, M., LEE, G., LELOUCHE, R., MEISSONNIER, V., DE VRIES, P. & ROCCATI, M. (1980). Projet RESEDA/1: Rapport sur les recherches effectuées du 1er octobre 1979 au 1er juillet 1980. *Rapport LISH/187*, Laboratoire d'Informatique pour les Sciences de l'Homme, Paris.

ZARRI, G. P., ORNATO, M., LEE, G., LELOUCHE, R., MEISSONNIER, V., DE VRIES, P., HAMMOND, L. & POMIAN, J. (1981). Projet RESEDA/1: Rapport sur les recherches effectuées du 1er juillet 1980 au 1er juillet 1981. *Rapport LISH/224*, Laboratoire d'Informatique pour les Sciences de l'Homme, Paris.

Distributed architecture and parallel non-directional search for knowledge-based cartographic feature extraction systems

Barbara A. Lambird, David Lavine

L.N.K. Corporation, Silver Spring, Maryland

and Laveen N. Kanal

Department of Computer Science, University of Maryland

Expert or knowledge-based system approaches are currently being viewed with great interest for their potential to handle the many difficult problems encountered in image understanding and cartographic feature extraction from remotely sensed imagery. This article presents an overview of the many types of knowledge that must be modeled in remote sensing and cartography, and discusses architectural and control aspects deemed important for cartographic expert systems. A distributed architecture and a control structure based on a parallel non-directional search algorithm are described and open problems are mentioned.

Introduction

The vast amount of incoming cartographic information makes it necessary to make greater use of available computer power. Due to the broad applications of remote sensing in critical areas such as crop monitoring, flood control, ice mapping in shipping lanes, weather forecasting, and pollution control such as oil spill containment, the throughput requirements for processing remotely sensed data are growing at a rapid rate. Yet manual photo-interpretation is a time-consuming, tedious, and error-prone task.

In the past, techniques for the automated or semi-automated extraction of cartographic features from remotely sensed images did not work very well. Many problems were encountered, including the sheer amount of information in each image, which taxed the memory and processing capabilities of the computers. The ambiguous and contradictory information that is present in the images makes image interpretation very difficult. Problems in dealing with perspective changes and the wide range of object scale and image resolution were encountered. In addition, there are many sources of geometric and radiometric variability which confound attempts at object detection. Finally, there has been a lack of adequate models relating physical principles to object appearance in images.

More recently, work on developing expert systems for image understanding and for cartographic feature extraction has sparked hope that these problems may be successfully handled. A well-known example of an image understanding expert system is ACRONYM developed by Brooks (1983). An examination of the strengths and limitations of this notable system points to the additional capabilities that will need to be provided in order to succeed in developing an expert system for cartographic feature extraction. Following a brief overview in section 2 of types of knowledge that must

221

be modeled in remote sensing and cartography, this article discusses the architectural and control considerations we deem important for cartographic expert systems. In section 3, we discuss why distributed problem solving and distributed architectures are suitable for this application. Section 4 suggests a possible control mechanism for handling a distributed expert system which appears to be particularly relevant in this context. Section 5 mentions several important open problems. Finally, section 6 presents concluding remarks.

2. Knowledge for feature extraction of remotely sensed imagery

The successful interpretation of remotely sensed images requires many types of knowledge. There is a great deal of knowledge particular to the remote sensing field. Many types of entities can appear in an image, and associated with each is a large collection of properties. Expert systems that interpret images need knowledge about image processing, image formation, object recognition and object modelling. Expert systems that need different types of knowledge should be able to handle multiple representations. In this section, we discuss some of the types of knowledge required for cartographic feature extraction.

REMOTE SENSING

First, knowledge about the external factors that induce variability in aerial imagery should be included. Automated or semi-automated recognition processes must be able to adjust to or account for variations that can be expected to appear in remote sensing and that are not caused by changes in the objects themselves. This knowledge derives from current sensor models, and models for geometric and radiometric distortion introduced in the image. We briefly mention some of these factors. More extensive discussions of this subject can be found in remote sensing texts such as Lillesand & Kiefer (1979) and Swain & Davis (1978).

Remote sensing is a method of gathering information about an object without any direct contact with the object. Most of the sensors used in imaging remote sensing applications are designed to respond to different parts of the electromagnetic spectrum. Some sensors, such as multispectral scanners, are passive where the sensors record ambient energy caused by solar and terrestrial radiation. Some sensors may be active, as in the case of radar systems where they provide the radiation which is reflected off the object.

The sun radiates mainly in the visible and infrared (IR) parts of the electromagnetic spectrum, and the earth radiates mainly in the thermal infrared (TIR) region. Not all of this radiation can be used by the remote sensor since particles and water vapor in the earth's atmosphere both scatter and absorb some of it. The overall effect of scattering is to decrease the amount of energy reaching the surface and to create "airlight" or background haze. The airlight can then enter the remote sensor and cause deterioration in the quality of the imagery. The effect of absorption is to decrease the amount of energy reaching the ground.

As a result of the scattering and absorption effects, only some regions of the spectrum reach the ground. This results in "atmospheric windows" (i.e. bands of the spectrum where transmission is approximately 70–100%) which are useful for remote sensing. These windows occur in the visible, near IR, middle IR, and thermal IR regions. The electromagnetic radiation which reaches the ground can then be reflected or thermally emitted by objects and this energy is then sensed by the remote sensors.

Models approximating the solar and terrestrial radiation, and the scattering and absorption effects represent some of the knowledge·that should be included in any cartographic expert system for remotely sensed imagery. In addition, temporary conditions such as haze, rain, and snow will also affect the appearance of objects. Knowledge of these effects should also be available to the expert system.

There are many types of remote sensors, e.g. photographic aerial cameras, multispectral line scanners (MSS), thermal infrared scanners (TIR), and side-looking airborne radars (SLAR). Each of these types of scanners and the distortion they introduce may be approximated by models. This also represents knowledge that must be incorporated in cartographic expert systems which interpret remotely sensed images.

The two basic types of distortion are radiometric distortion and geometric distortion. Radiometric distortion occurs when the intensity of the radiation received by the sensor undergoes changes that are not directly caused by the objects. The radiation which is measured by the remote sensors undergoes changes through absorption and reflection processes as discussed above. Different conditions, such as the amount of dust or water vapor in the atmosphere, changes in the sun–object–sensor angle, and the type of surface of the objects, can greatly affect the appearance of the objects. In addition, the response of each detector in a sensor is different from any other detector. All these effects cause radiometric distortions.

In addition to these radiometric effects, the images are affected by geometric distortions which can also greatly change the appearance of the objects. For example in some scanner imagery, straight roads will become "S" shaped. The sensor geometry introduces geometric distortion which changes the physical appearance of an object. Even when sensor-based geometric distortions are partly corrected, other geometric distortions present can not always be removed. The basic types of geometric distortions common to all remote sensors are:

(1) distortions in the image introduced by the topography, i.e. by the varying elevation of the terrain;
(2) distortions caused by the sensing mechanism;
(3) distortions caused by the recording mechanism; and
(4) distortions caused by a non-ideal flight path of the aircraft carrying the sensor.

Some types of aerial images have additional distortions: at high altitudes both the curvature of the earth and the motion of the earth become noticeable. Clearly an expert system for cartographic feature extraction should have access to models of complex phenomena, such as the above types of distortion, not all of which are fully developed at present.

OBJECTS

Knowledge about the objects which are likely to be encountered must be included in the expert system. This information may include shape, structural, material composition and surface properties of the objects. Three-dimensional models of the object may be required. Relational properties of the object may also be necessary. For example, the knowledge that a bridge joins two bodies of land across a body of water or of land of lower elevation can be of great use in locating bridges. Figure 1 shows a simple example

ATTRIBUTES:
  ROAD (REGION): PROP-TABLE-PTR
    WHERE NOT (CLASSIFICATION (PROP-TABLE-PTR)
      ∈ {VEGETATION, WATER})
    AND ELONGATED (PROP-TABLE-PTR).

  LAND (REGION): PROP-TABLE-PTR
    WHERE CLASSIFICATION (PROP-TABLE-PTR) = VEGETATION
    AND NOT (HOMOGENEOUS (PROP-TABLE-PTR))
    AND AREA (PROP-TABLE-PTR) > LARGE.

  WATER (REGION): PROP-TABLE-PTR
    WHERE CLASSIFICATION (PROP-TABLE-PTR) = WATER.

RULES:
  BRIDGE
    IF rg(1) = ROAD
      AND rg(1) NEIGHBOR OF rg(2)
      AND rg(2) = WATER
      AND rg(3) NEIGHBOR OF rg(1)
      AND rg(3) = WATER
      AND NOT (rg(2) NEIGHBOR OF rg(3))
      AND rg(4) NEIGHBOR OF rg(1)
      AND rg(4) = LAND
      AND rg(4) NEIGHBOR OF rg(2)
      AND rg(5) NEIGHBOR OF rg(1)
      AND rg(5) = LAND
      AND rg(5) NEIGHBOR OF rg(2)
    THEN BRIDGE = rg(1)
      AND USER CHECK-BRIDGE-ELEVATION ((REGION): rg(1),
        (REGION): rg(2), (REGION): rg(3), (REGION): rg(4),
        (REGION): rg(5)).

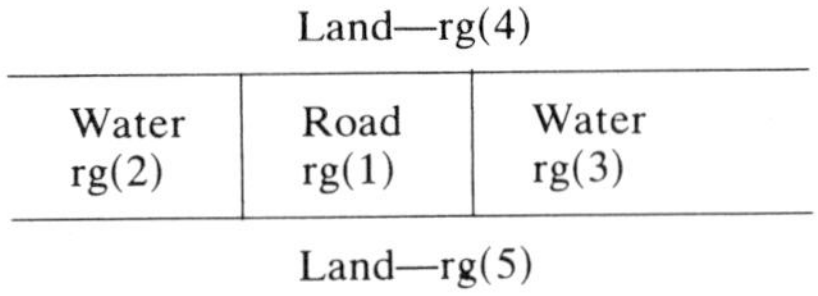

Land—rg(4)

Water rg(2)	Road rg(1)	Water rg(3)

Land—rg(5)

Fig. 1. A bridge first erroneously classified as a road and then correctly recognized by the expert system's use of context. Subsequently this information was used to correct the incorrect elevation data.

of how elevation may be used to disambiguate otherwise ambiguous objects, in this case a road from a bridge.

Further properties of the object could describe their temporal variation or their interaction with the environment. Crops change radically in appearance throughout the year. Soil reflectance can change drastically with variations in the moisture content. Functional properties of the objects can be greatly useful in recognition and should be included. For example, automobiles are used for transportation and thus they are likely to be found near roads.

The physical structure of objects can be used to predict their appearance in various types of remotely sensed images. This knowledge should be provided and should include methods for determining radiometric response characteristics. Principles of geometrical reasoning should also be included. An example of geometric reasoning is the effect of perspective in viewing three-dimensional shapes.

IMAGE PROCESSING

Image processing knowledge must obviously be represented in the expert system. This knowledge should include description of a wide range of image processing techniques, their uses, and the information required to use them. Information about the image formation and acquisition processes should be available. In some imagery the original data may not lie on a rectangular grid, so various interpolation processes may be necessary to transform the data. This is true for several widely used data sources such as Landsat and Seasat data. The transformations will affect the accuracy of the image, which may be of great importance if high precision measurements are required. In general, the expert system should know the image scale and the characteristics of the geometric and radiometric distortion resulting from the sensor. Since a wide range of object sizes are present in remotely sensed images, knowledge of how to adapt the image processing techniques to objects ranging widely in size should be included. Varying image resolutions give rise to similar problems and must be dealt with.

OBJECT RECOGNITION AND MODELLING

Finally, a cartographic expert system should include knowledge about object recognition and object modelling. Object recognition is the process of recognizing an instance of an object given a model of it. Object modelling is the process of separating an object from the background, given generic descriptions of relations between object components. An example of the object modelling process is Waltz's (1975) work on corner labelling to find objects in an image.

3. Distributed problem-solving

Distributed problem-solving (DPS) is a basic method of attack for solving problems which are too complex to be solved by a single processor: the problems may require too much information or too much processing power. The knowledge sources needed for cartographic expert systems are so varied and the data so voluminous that this approach appears most promising. In this section, we discuss the nature of distributed problem-solving and suggest a possible architecture for distributed cartographic expert systems.

NATURE OF DPS

In distributed problem-solving (DPS), a complex problem is subdivided into a set of distinct subproblems which are easier to solve. Each subproblem can then be assigned to a processor for solution. DPS has several advantages (Chandrasekaran, 1981): each processor need only solve a more limited problem. The total input to the limited problem should be correspondingly smaller. Distributing the problem among several processors allows parallel processing to take place. In cases which require real-time processing, parallel processing may be absolutely necessary in order to accomplish the tasks. If multiple processors are used then parts of the system can fail or degrade but still allow partial results to be obtained, or at least the cause of failure to be determined.

Finally, distributed problem-solving allows a modular structure which can be easier to expand or be more adaptable to change, if properly constructed.

There are basically two architectures for distributed problem-solving: network and hierarchy. In a network of processors, each processor can communicate directly with any other processor. While this architecture facilitates exchange of information, it requires that all processors must have the information processing capability of the processor with the largest information load. Thus this architecture involves a trade-off of increased ease of inter-processor communication against a requirement that all processors must be complex enough to handle the greatest information load.

In a hierarchy of processors, the lines of direct communication are limited to be between only those processors which are directly connected. Thus, some processors can not directly communicate, but must go through intermediary processors. In this case, most processors need no longer be capable of handling the greatest information load. However, since information may now be passed through intermediary processors, "filtering" or "biasing" of the information will occur. This biasing may be good or bad, depending on the application. For example, suppose a processor inquires through the hierarchy whether a detected feature on an unrectified line scanner image is a straight road. As the inquiry is passed through the hierarchy, the line scanner expert would appropriately change this inquiry since straight lines are distorted into variously shaped curves depending on the orientation of the line with respect to the line scanner. In this example, the biasing or filtering of the information is necessary. However, the hierarchical architecture requires that the hierarchical structure be constructed. Thus, this architecture involves a trade-off between decreased ease of inter-processor communication with the added complexities of information filtering, and a lessening of the requirement that all processors must be complex enough to handle the greatest information load.

In the past, most expert systems have distributed the knowledge (for example, into discrete rules, frames, etc.) but have kept the control or processing burden on a highly centralized controller. For applications with a very high information and processing load, such as all image understanding expert systems, the burden rapidly taxes the capabilities of most systems. In order to construct useful image understanding expert systems, the processing burden will have to be distributed. This means new expert system architectures need to be developed. Later parts of this article describe a general distributed architecture for expert systems and a method for allowing parallel processing control.

DISTRIBUTED EXPERT SYSTEMS

Chandrasekaran (1983) has suggested that expert systems can be organized as a "cooperating community of specialists". In this case, the knowledge is divided among a set of structures each of which utilizes the most appropriate type of problem-solving method. Each structure can then be decomposed into a hierarchy of specialists which share the same type of problem-solving method. In this architecture, knowledge is not separated from the control mechanism but is imbedded in it. Thus, the problem-solving is distributed throughout the expert system.

Each specialist contains its own knowledge base and corresponding inference mechanism. Usually specialists high in the hierarchy are more general, while specialists lower in the hierarchy are more specific. For example, in a cartographic expert system a higher-level specialist could be a specialist in interpreting urban areas, while a lower-level specialist could be a specialist in the recognition of automobiles. Communication between specialists can occur readily along the lines of the hierarchy. Communication between specialists not directly connected by the hierarchy can be accomplished either through the hierarchy or through an external blackboard. The blackboard contains the status of the system, i.e. which specialists have been explored and their status.

The distributed expert system has several advantages. First, it allows different types of knowledge representation and the appropriate problem solving methods to be included in one system. This advantage is extremely important for cartographic expert systems, since there is a large amount of very different kinds of knowledge needed for the interpretation of remotely sensed images. The requirement of only one knowledge representation and corresponding problem-solving method is too restrictive. This subject is explored in more detail in the next section. Much of the control and "focus of attention" problems are alleviated since control can only pass along the lines of the hierarchy. The distributed architecture does have disadvantages. The domain knowledge must be carefully structured in a hierarchy. In addition, in a large complex system communication through the blackboard may be difficult to implement efficiently.

ACRONYM

ACRONYM (Brooks, 1983) is an image understanding expert system that illustrates some of these ideas. ACRONYM is one of the most sophisticated expert systems from a number of points of view. It does symbolic reasoning on two-dimensional images using three-dimensional models. What is of interest, in this context, is that it incorporates three separate expert systems, each of which have their own knowledge representation and type of reasoning. The three expert systems co-operate to interpret the images. Very briefly, the expert systems are: (1) a "prediction" system that uses the three-dimensional models to predict geometrically invariant features to look for in the image. This system uses geometrical reasoning. (2) A "description" system that uses the images to get descriptions of possible image features. This system uses image formation reasoning. (3) An "interpretation" system that uses the descriptions from the second system to find constraints and check the consistency of the results. This system uses graph matching to perform its reasoning. The three systems are iterated (prediction–description–interpretation) in order to get increasingly more detailed interpretations of the image.

In addition to the co-operating expert systems, ACRONYM does include some of the knowledge that was discussed in section 2. ACRONYM includes some knowledge about object recognition, object modelling, image formation, sensors, and illumination. However, most of this knowledge is minimal. ACRONYM has a sophisticated modelling system for three-dimensional objects, but the objects must be rigid or be composed of rigid components. In addition, only limited object relations and properties are handled by ACRONYM. Thus, while ACRONYM represents a significant step in the inclusion of knowledge, many additional features must be added in order to incorporate knowledge of the type described in section 2.

The three expert systems in ACRONYM are all rule-based productions systems which do not explicitly relate the knowledge in a hierarchy of specialists, and must be separately controlled. The focus-of-attention problem mentioned above could lead to difficulties. As reported in the literature, ACRONYM has only been tested on a limited amount of imagery for a small set of objects. The control mechanisms used in ACRONYM are unlikely to be able to handle interpretation of complex images containing a large set of complex objects. For the cartographic feature extraction problem, more general control structures should be considered. In the next section, we briefly describe a control approach which we have been investigating because we feel it is more suited to this domain.

4. Control

Using an expert system to extract cartographic features from an image may be thought of as a problem in search. The requirement that the search be distributed among a set of processors gives rise to the need for a parallel search algorithm. This algorithm should require communication of small amounts of information at infrequent intervals since each expert module will require a fairly sophisticated processor and the intimate communication among many such processors is time-consuming and prone to failure.

Further requirements are imposed on the search method by the cartographic problem. Due to the large amount of information present in images, expectations as to the probable contents of the scene should be used to direct the search for primitive features. These expectations should be computed during the course of the scene analysis. For example, the knowledge that a large body of water is likely to be present in an image may trigger the application of certain texture measures to locate the body of water. On the other hand, there are many features which may occur in a given image even though there is no *a priori* means of assessing the likelihood of their occurrence. In such situations, primitive image processing operators, such as edge or texture detectors, may be applied to the image to uncover the possible presence of higher-level structures, such as buildings. This "bottom-up" search can be very time-consuming since it leads to a large number of hypotheses, most of which are wrong. As soon as the presence of such a higher-level structure is suspected, it may be reasonable to resume a top-down approach, where high-level knowledge can be used to guide the search. A search algorithm used to control the cartographic expert system should be capable of moving between top-down and bottom-up modes of search depending on the characteristics of the data already found.

The next two sections describe a parallel search procedure capable of both top-down and bottom-up modes of search. This procedure was originally developed (Stockman, 1979) to guide search in the analysis of medical waveforms using methods from syntactic pattern recognition. The underlying search procedure, known as SSS*, has been the subject of considerable analysis (Stockman, 1979; Roizen & Pearl, 1983; Campbell, 1981). The waveform analysis package, known as WAPSYS, which contains SSS* has proven to be a fast, flexible means for the analysis of several types of medical waveforms (Stockman & Kanal, 1983; Xiong, Lambird & Kanal, 1984). The third section discusses two aspects of the dynamic control that will be needed in the cartographic expert system.

SSS*

The SSS* algorithm is a best-first state space search procedure developed for application in structural pattern recognition. This algorithm was designed to provide flexibility in search by allowing both model-directed and data-directed search. In particular, the following design criteria were used.

(1) Ambiguous interpretations should be allowed and developed in a best-first manner.

(2) *A priori* knowledge of the problem domain should be available to make hypotheses about the data for subsequent verification.

(3) Key events in the data, as detected by low-level interrogation, should be capable of triggering a search for higher-level structures in the data.

(4) The order of application of (2) and (3) should be determined dynamically, based on some optimality criterion.

In addition to the above features, SSS∗ is readily adaptable to distributed problem solving.

SSS∗ uses a problem reduction representation (PRR), AND/OR graph, or a grammar data base. Structures of interest are decomposed in terms of alternative (OR) and component (AND) substructures. Structures which cannot be further decomposed are regarded as primitives. In the cartographic context, the AND/OR graph consists of modules describing features such as roads, buildings, forests or towns, or methods for recognizing cartographic features. The descriptive components may include a variety of descriptors, such as the geometric layout of an object, texture properties, and spectral features. Primitives are detected by low-level image processing operations or by interrogating a human user of the system. The large number of primitive detectors applicable to a given image necessitate the use of a good control strategy.

SSS∗ controls the search by matching the PRR model to data. The algorithm terminates, if possible, with a description of the structure in question. This description takes the form of a state tree for the structure. Components of the tree are labeled and parameter values are present when appropriate.

A key feature in the efficiency of the search algorithm is the ordering of components based on the expected ease of detection. This ordering may be based on experimental studies or subjective assessment. By examining the components using this ordering, inappropriate branches of the AND/OR graph can be rapidly eliminated. For example, the component which is easiest to detect may be a large rectangular structure. This component was listed as best in the ordering since it was the easiest to detect. If this structure is not found, then the more time-consuming search for one of the other components need not be performed.

Matching of the PRR to data is accomplished by applying a sequence of operators which generate a sequence of states representing partial state trees. Both top-down and bottom-up operators may be applied, depending on whether a structural goal has been set or a structure has been recognized. Partial state trees are stored in a linear encoding and rated as to the quality of the match with the data. As alternative OR goal structures are generated, WAPSYS creates competing trees. This parallel development of competing trees allows discovery of multiple solutions to a problem.

WAPSYS uses the A∗ search algorithm defined in Nilsson (1980) to control generation of the states. A global data base known as the State Space Representation (SSR) contains an encoding of the states which have been generated. Initially, the only states in the SSR are structural goals for top-down analysis and primitive structural goals for bottom-up analysis. Each state is assigned a merit which is an estimate of how far the state is from a goal state. The A* algorithm expands the state which has the highest merit value and places the expanded state in the SSR. Currently WAPSYS uses the minimum value of any recognized primitive structure in a state as the value of the merit. This approach of defining the quality of recognition of an object as being the merit of the most poorly detected part leads to some difficulties in search. One moderately bad, but acceptable, evaluation will greatly reduce the chance of the state being expanded until other states which are, on the average, worse have been expanded. Various schemes to overcome this limitation have been devised using some type of average error, but they are not guaranteed to arrive at an optimal solution.

The operators used in the A* state space search provide the ability to incorporate both top-down and bottom-up search. These operators, are described in more detail in Stockman & Kanal (1983). Recognized structures may have associated attributes. These attributes may contain information such as the physical location of the recognized structure. These attributes can be used to constrain later parts of the search.

The primary application of WAPSYS has been to the analysis of waveforms, though the paradigm has a much broader range of applications. Through its ability to combine top-down and bottom-up search and its ability to develop competing solutions in parallel, WAPSYS promises to provide a flexible control structure for the problem of distributed expert system processing.

PARALLEL NON-SEQUENTIAL SEARCH

A parallel implementation of the SSS* search algorithm has been developed in the context of Branch and Bound procedures (Kanal & Kumar, 1981; Kumar & Kanal, 1983, 1984). The search space is partitioned and each part is searched in a depth-first search. Each time alternate structural models arise in the AND/OR graph search, the AND/OR subtrees corresponding to the alternatives, can be sent to separate processors. Using this approach, the procedure can be operated as a set of processors working independently and asynchronously. Only the merit of the best partial parse tree encountered in the search so far needs to be communicated among the processors.

Parallel implementations of Branch and Bound (B&B) search algorithms have been previously developed and studied (El-Dessouki & Huen, 1980; Imai, 1979; Mitten, 1964). A parallel depth-first B&B algorithm in which n concurrent processors work on the n "deepest" (or most recently generated) subproblems was proposed by Imai (1979). All generated subproblems are kept in a common memory accessible by all the processors. Ma & Wah (1981) propose a parallel B&B algorithm in which n processors work on the n "best" subproblems. Both of these implementations require a high level of communication between processors or between processors and a common memory. These communication requirements cannot easily be met with general purpose, loosely coupled distributed architectures.

We now briefly describe the general framework for parallel search and a parallel implementation of SSS* developed in Kanal & Kumar (1981) and Kumar & Kanal (1984). The basic approach to B&B problems is to decompose the search space into disjoint areas and search each area in a depth-first fashion. These searches can be performed concurrently. Each time a processor encounters a better solution, it informs all other processes in the system of the merit. Whenever a partial solution is encountered which cannot be better than the current global optimum, that solution is discarded. Since a processor only uses shared information to decide if it should give up a partial solution and examine another part of the search space, it never has to wait for input from another processor. This feature is highly desirable if the algorithm is to be implemented on a loosely coupled architecture on which interprocessor communication time is slow. Evaluation of the speed-up achievable with this parallel algorithm is difficult to evaluate on theoretical grounds, since it depends largely on how well the search space can be divided into areas that require approximately equal search time.

A solution tree T of an AND/OR tree G, is usually defined as a sub-tree such that (a) the root node of G is the root node of T and (b) if a non-terminal node of G is in T, then all of its immediate successors are in T if they are of type AND, and exactly one of its immediate successors is in T if they are of type OR. Complementary to this "AND" solution tree formulation, an "OR" solution tree may be defined as follows (Kanal & Kumar, 1981): (a) the root node of G is the root node of solution tree T and (b) if a non-terminal node of G is in T then all of its immediate successors are in T if they are of type OR, and exactly one of its immediate successors is in T if they are of type AND.

The parallel implementation of the B&B algorithm requires depth-first search in the AND solution tree space and best-first search in the OR tree search space. Unfortunately, sequential SSS* does the reverse of this. To overcome this problem, a dual version of SSS* has been formulated, in which partial OR solution trees are searched in best-first order. This approach called dual-SS* is a depth-first search in the AND solution tree space. Dual-SS* keeps track of the best complete AND-solution tree found so far. Any partially explored AND-solution tree which has a merit lower than the current best solution is eliminated. Furthermore, the current best solution tree is replaced only when a solution tree with higher merit is encountered. Thus the dual-SS* algorithm can be used for the search in the parallel SSS* implementation.

Although the distributed architecture described in this section seems most suited to cartographic applications, much research needs to be done and many problems to be solved before a true cartographic expert system can be realized. We discuss some of these problems in the following section.

DYNAMIC CONTROL

The advent of distributed expert systems naturally gave rise to the possibility of using a loosely coupled network of processors. While the assignment of different expert modules to different processors may appear to be a natural means for assigning software to processors, this can lead to serious difficulties. Expert modules differ considerably in their system resource requirements on both a static and dynamic level. From the static point of view, the amount of code occupied by different expert modules will vary. In addition, the *a priori* expected usage of these modules will also vary. On the dynamic side, the frequency of usage of certain modules will depend heavily on the image being processed at a given time. Ideally, run-time assessments of images under analysis should be performed to aid in the decision as to how to distribute the modules. A simpler, but still nontrivial approach is to assume the modules are assigned to fixed processors and focus on the problem of controlling the flow of information.

A second type of dynamic control lies in the run-time construction of the AND/OR graphs. As discussed in earlier sections, the distributed problem solving system consists of a hierarchy of specialists. When a specialist is invoked, it must decide which if any of its sub-specialists it must call upon to help solve its problem. In cartography, the applications are too complex to enumerate all the possible combinations of specialists that would be needed to solve all possible problems. For example, towns have such an enormous number of variations that it is not feasible to have an explicit AND/OR graph to capture these possibilities. As an alternative, the structure of the AND/OR graph can be dynamically constructed during the search process. The selection of specialists will be based on the static structure of the underlying tree of the expert system and the data observed so far in the search.

As an example, let one of the specialists in the cartographic expert system be a specialist on the recognition of urban areas. Since there are many different types of urban areas (for example: heavy industry, light industry, suburban, etc.), there will probably be sub-specialists which specialize in the recognition of these different types. Thus, when the urban area specialist is invoked, it may have enough information to realize the area is either a light industrial area or a suburban community. In this case, the dynamic AND/OR graph need only include those two sub-specialists in an OR configuration. Depending on the data observed in the search so far, the light industrial urban area specialist may invoke experts to seek railroads, waterways, large building complexes, and evidence of pollution. The suburban specialist may institute a search for road networks and large numbers of small buildings in highly structured patterns. These two specialists would invoke their sub-specialists in an AND configuration.

5. Some problems to be solved

In this section, we present several problems that are illustrative of the type of problems that need to be solved before practical cartographic expert systems can be effectively developed and implemented.

A fundamental problem in cartographic expert systems is the resolution of contradictory information resulting from the use of several types of sensors. In the early stages of processing, the different sensed images may be handled independently. Radar, multispectral, and infrared imagery are but a few of the sources of pictorial information available. Elevation matrices and maps provide additional knowledge useful for image interpretation. While the tremendous amount of information available from these sources offers hope for considerable improvement in the reliability of scene analysis, many obstacles must be overcome.

Complex problems in interpreting information from multiple sensors are common in scene analysis. For example, long narrow features, such as roads, have been used to compute the transformation needed to register (align) two images of the same area. Preliminary feature analysis may indicate the presence of roads in the two images, but due to various types of distortion, the shapes of the roads may vary enough to make accurate registration difficult. We have encountered this problem in work on registering radar and optical images. A simple weighting of the beliefs in the reliability of the two sources of information can lead to estimated road positions which are seriously inconsistent with both sources. In this type of problem, the expert system is required to develop geometric descriptions satisfying a variety of quantitative and qualitative constraints.

Once an image is registered to other images, maps, and elevation matrices, search for other roads can make use of the multiple information sources. Line or edge detectors in high resolution monochromatic imagery can be used to locate prospective sites for roads. Elevation information obtained from stereo imagery imposes further constraints on possible road locations. Multispectral imagery can be used to check for road surface materials. Rules limiting the curving of roads in some locales may be used to limit further possible road detections.

A major problem that is inherent in developing any distributed expert system that would use parallel processing is the problem of assignment of expert modules to particular processors. A successful solution to this problem will require procedures for predicting the relevance of expert modules and the complexity of their search problems during the analysis of a scene. These prediction procedures could initially be based on some simple measures of search complexity, such as some measures of information content in an image (such as average edge density or texture complexity in sample areas). The problem is made even more difficult by the dynamic structuring of the AND/OR graph. This problem of predicting the relevance of expert modules for the purpose of processor allocation is a virtually unexplored area.

6. Concluding remarks

We have described the complex types of knowledge that will have to be available to a cartographic feature extraction expert system. We have described the distributed architecture and a control structure based upon a parallel non-directional search algorithm which we consider highly suited to this application. We have also mentioned a number of problems which remain to be solved before practical systems can be realized. Many aspects of cartographic expert systems have not been touched on in this article. Some of these, for example, the role of low-level statistical processing and classification integrated with the higher-level expert system are described in our previous reports (Lambird, Lavine & Kanal, 1981). We have also not touched upon methodologies for processing contradictory or ambiguous evidence which is an important and challenging problem in itself. We are continuing to explore the ideas presented in this article and implementing some of them in some of our current projects.

This research was supported in part by L.N.K. Corporation, 302 Notley Ct, Silver Spring, Maryland 20904, U.S.A. and in part by grant No. MCS-81-17391 from the Mathematical and Computer Sciences Division, N.S.F. to the Machine Intelligence and Pattern Analysis Laboratory in the Department of Computer Science, University of Maryland, College Park.

References

BROOKS, R. A. (1983). Model-based three-dimensional interpretations of two-dimensional images. *IEEE Transactions on Pattern Analysis and Machine Intelligence*, **PAMI-5**, 140–150.
CAMPBELL, M. (1981). Algorithms for the parallel search of game trees. *Technical Report 81–8*, Department of Computer Science, University of Alberta, Edmonton.
CHANDRASEKARAN, B. (1981). Natural and social system metaphors for distributed problem solving: Introduction to the issue. *IEEE Transactions on Systems, Man, and Cybernetics*, **SMC-11**, 1–4.
CHANDRASEKARAN, B. (1983). Expert systems: Matching techniques to tasks. *New York University Symposium on Artificial Intelligence Applications for Business*, 18–20 May 1983.
EL-DESSOUKI, O. I. & HUEN, W. H. (1980). Distributed enumeration on network computers. *IEEE Transactions on Computers*, **C-29**, 815–825.

IMAI, M., YOSHIDA, Y. & FUKUMURA, T. (1979). A parallel searching scheme for multiprocessor systems and its application to combinatorial problems. *Proceedings Sixth International Joint Conference on Artificial Intelligence*, pp. 416–418.

KANAL, L. N. & KUMAR, V. (1981). Parallel implementations of a structural analysis algorithm. *Proceedings IEEE Computing Society Conference on Pattern Recognition and Image Processing*, Dallas, pp. 452–458.

KUMAR, V. & KANAL, L. N. (1983). A general Branch and Bound formulation for understanding and synthesizing AND/OR tree search procedures. *Artificial Intelligence*, **21**, 179–198.

KUMAR, V. & KANAL, L. N. (1984). Parallel Branch and Bound formulations for AND/OR tree search. *IEEE Transactions on Pattern Analysis and Machine Intelligence* (to appear).

LAMBIRD, B. A., LAVINE, D. & KANAL. L. N. (1981). Interactive knowledge-based cartographic feature extraction. *ETL-0273*, U.S. Army Engineer Topographic Laboratory, Fort Belvoir, Virginia 22060, U.S.A. (October).

LILLESAND, T. & KIEFER, R. (1979). *Remote Sensing and Image Interpretation.* New York: John Wiley & Sons.

MA, E. & WAH, B. W. (1981). MANIP—A parallel computer system for solving NP-complete problems. *Proceedings N.C.C. AFIPS Conference.*

MITTEN, L. G. (1964). Composition principles for synthesis of optimal multi-stage processes. *Operations Research*, **12**, 610–619.

NILSSON, N. (1980). *Problem-solving Methods in Artificial Intelligence.* Palo Alto: Tioga Publishing Co.

ROIZEN, I. & PEARL, J. (1983). A minimax algorithm better than alpha-beta? Yes and no. *Artificial Intelligence*, **21**, 199–220.

STOCKMAN, G. C. (1979). A minimax algorithm better than alpha-beta? *Artificial Intelligence*, **12**, 179–196.

STOCKMAN, G. C. & KANAL, L. N. (1983). Problem reduction representation for the linguistic analysis of waveforms. *IEEE Transactions on Pattern Analysis and Machine Intelligence*, **PAMI-5**, 287–298.

SWAIN, P. & DAVIS, S. (1978). *Remote Sensing—The Quantitative Approach.* New York: McGraw-Hill.

WALTZ, D. (1975). Understanding line drawings of scenes with shadows. In WINSTON, P. H., Ed., *The Psychology of Computer Vision.* New York: McGraw-Hill Book Co.

XIONG, F. L., LAMBIRD, B. A. & KANAL, L. N. (1984). An experiment in the recognition of electrocardiograms using a structural analysis algorithm. *Proceedings IEEE 1983 International Conference on Systems, Man, and Cybernetics*, Bombay and New Delhi, India, December 1983.

A model for the interpretation of verbal predictions

ALF C. ZIMMER

Department of Psychology, Westfälische Wilhelms Universität, Münster, Federal Republic of Germany

There is a marked gap between the demands on forecasting and the results that numerical forecasting techniques usually can provide. It is suggested that this gap can be closed by the implementation of experts' qualitative predictions into numerical forecasting systems. A formal analysis of these predictions can then be integrated into quantitative forecasts.

In the framework of possibility theory, a model is developed which accounts for the verbal judgments in situations where predictions are made or knowledge is updated in the light of new information. The model translates verbal expressions into elastic constraints on a numerical scale. This numerical interpretation of qualitative judgments can then be implemented into numerical forecasting procedures.

The applicability of this model was tested experimentally. The results indicate that the numerical predictions from the model agree well with the actual judgments and the evaluation behavior of the subjects.

The applicability of this model is demonstrated in a study where bank clerks had to predict exchange rates. The analysis of qualitative judgments according to this model provided significantly more information than numerical predictions.

A general framework for an interactive forecasting systems is suggested for further developments.

Introduction

In the analysis of human predictive behavior the behavioral scientist is confronted with a puzzling situation: predicting future events correctly is undoubtedly of high survival value for any species, especially for humans living in complex interactions with their material and societal environment. This has led Friedman & Willis (1981) to assume that prediction can be regarded as the master motive in human behavior. However, a host of studies has seemingly proved that people are rather inefficient in taking into account the information necessary for a correct prediction (Meehl, 1954; Hogarth & Makridakis, 1981).

A closer analysis of the studies casting doubts on the human efficiency in forecasting reveals that the superiority of statistical predictions is especially prominent if the data are numerical and if the requested prediction is also numerical. On the other hand, if the data are non-numerical and highly configural (Lindzey, 1965), subjective judgments can be better than statistical predictions. Furthermore, experts are able to handle up to 10 variables simultaneously (Phelps & Shanteau, 1978). This result indicates that the limitations of capacity in human information processing depend on the amount of expertise or, in general, on the amount of elaboration of the applied internal model.

A tentative solution for the described contradictory situation might be that humans are comparatively efficient in qualitative forecasting but quite unsuccessful in quantitative prediction [one notable exception is the case of meteorologists, see Murphy & Winkler (1975)]. If this is the case, forecasting and planning tasks which permit verbal,

DEVELOPMENTS IN EXPERT SYSTEMS
ISBN 0-12-187580-6

qualitative information processing can be assumed to be less flawed by information processing biases as listed by Hogarth & Makridakis (1981, pp. 117–120).

For diagnostic tasks the necessity to implement subjective evaluation into the processing of "hard" quantitative information has led to the development of expert systems, for instance, MYCIN (Shortliffe, 1976) in which the heuristics of medical experts are represented as production rules. These production rules combined with the computers' efficiency in storing and accessing data have made MYCIN successful, and have triggered the development of other expert systems for different areas of expertise (e.g. organic chemistry, DENDRAL; geological survey, PROSPECTOR; systems programming, R1). The success of expert systems suggests to look for the heuristics and rules underlying the qualitative predictions of experts and to combine them with the results of quantitative forecasts. A necessary prerequisite for such an integrated forecasting system is the formal description of the rules underlying subjective predictions and the development of an algorithm which translates qualitative judgments into parameters or intervals in $\mathbb{R}$. If such a system turns out to work successfully, "qualitative projection techniques" and "quantitative projection techniques" (Cleary & Levenbach, 1982, p. 6) are no longer mutually exclusive alternatives but two mutually supportive facets of forecasting.

The description of the forecasting process by Butler, Kavesh & Platt (1974) captures very well the tenets of such an integrated system:

> In actual application of the scientific approaches, judgement plays, and will undoubtedly always play, an important role .... The users of econometric models have come to realize that their models can only be relied upon to provide a first approximation—a set of consistent forecasts which then must be "massaged" with intuition and good judgment to take into account those influences on economic activity for which history is a poor guide. (p. 7.)

This integrated system of qualitative and quantitative techniques in forecasting might also serve an additional function, that is, increasing the acceptability of forecasts. Among professional forecasters the following paradox of forecasting (better, perhaps, dilemma) is reported: either the result of the forecast is in line with the intuitions of the customer, then it is taken into account, or the result of the forecast and the intuitions clash, in which case the forecast is discarded. In both cases nothing is done which had not been done without the results of the forecast. By implementing the customer's intuitions (e.g. his or her "subjective econometric model") it can be expected that the additional information of the numerical forecast becomes more acceptable for the customer because it is computed in relation and addition to the analysis of the plausible reasoning of the customer or other experts.

Furthermore, under certain conditions the implementation of subjective qualitative judgments is necessary to provide an adequate database for predicting complex variables such as, for instance, productivity. The importance of productivity for economical growth is known and almost everyone has a notion about the meaning of the term productivity. If, however, numerical forecasting methods are applied to productivity only measurable variables (e.g. units fabricated during a time period in a production line) can be analyzed. Other important variables of productivity (e.g. creativity in design or planning, effectivity in administration, control of turn-over) have to be discarded, because they cannot be stated numerically.

The claim is made that the knowledge of experts in its greater part usually consists of such qualitative variables stated verbally. Since verbal statements do not fit into numerical models of forecasting they are usually neglected in the application of these models.

From a psychological point of view the investigation of predictive behavior has the following objectives (cf. Kahnemann & Tversky; 1973):

(i) How valid are human predictions?
(ii) What are the conditions for making valid or invalid predictions?
(iii) What makes predictions credible?

In the context of this paper I will confine myself mostly to point (ii) and only at the end will I touch point (i). Point (iii) is mainly dealt with in the context of research on persuasion and will not be discussed here except for the claim that by implementing the addressee's expertise into the prediction he or she will probably be more willing to take the prediction seriously.

In the context of cognitive psychology, what has been analyzed mostly are the conditions underlying the lack of validity in predictions made by human subjects (e.g. Edwards, 1968; Kahnemann & Tversky, 1973). A number of persistent biases have been shown to influence the validity of predictions negatively:

(i) conservatism, i.e. failing to take into account new information;
(ii) availability, i.e. giving too much emphasis to pieces of information easily accessible in one's own memory; and
(iii) representativeness, i.e. disregarding the statistical properties of information (e.g. sample size, correlation, base rate, randomness).

For an overview of research on these biases see Kahnemann, Slovic & Tversky (1982), Nisbett & Ross (1980), or Hogarth & Makridakis (1981). Throughout the studies cited above it is stressed that the biases found are due to the mistaken application of heuristics. Heuristics are tools for the mind that are usually helpful for tackling the complexities of the environment of humans (see, for example, Lenat, 1982). What is missing in these studies is—with few exceptions—the analysis of conditions under which these heuristics are valid and the development of means which make their application possible even in situations which are more complex than the ones under which they have emerged originally (see Sjöberg, 1982; Zimmer, submitted).

One reason for the observed lack of validity for predictions in the studies reported above might be that in many of the experiments subjects were forced to give numerical estimates of their confidence, their predictions, or their subjective probabilities for further events. As argued elsewhere (Zimmer, 1983), it seems plausible to assume that the usual way humans process information for predictions is similar to putting forward arguments and not to computing parameters. Therefore, if one forces people to give numerical estimates, one forces them to operate in a "mode" which requires "more mental effort" and is therefore more prone to interference with biasing tendencies. One main difficulty in analyzing verbal information processing is that the meaning of verbal expressions is vague and therefore cannot be translated easily into formal (e.g. numerical) terms. Fuzzy set theory provides comparatively simple methods to account for vagueness and, by putting elastic numerically stated constraints on vague concepts, fuzzy-set models of human judgment (Zimmer, 1980) permit the translation of verbal expressions into numerical expressions.

A formal description of knowledge underlying predictions

From the point of view of human information processing the framework for people's world knowledge can be summarized as in Table 1. The application of available procedures to information stored in the appropriate modes of representation generates new knowledge which is not due to observations. What is especially of importance in the context of prediction are the conclusions which can be drawn from propositionally represented knowledge by means of the rules of logic or the rules of an argumentative discourse. Such conclusions can be used in hindsight for explanations and in foresight for predictions. The application of the rules of logic implies that the meaning of the propositions is unambiguous, and for most systems of logic it is furthermore necessary that the propositions be either true or false. Whenever these conditions are given it can be decided for any deduction if it is valid or not. But if the meaning of the propositions and their truth-values are vague, then the rules of classical two-valued logic are not applicable. Furthermore, it is questionable if formal logic is an apt tool for modelling human reasoning. Among others, Begg (1982) has shown that people in solving syllogisms do not apply the rules of logic but "play the language

TABLE 1
Human information processing depends on

---

(1) Modes of representing
    (i) propositional
               linguistic
                   on the word level
                   on the sentence level
                   on the story or script level
               numerical
               general symbolic
    (ii) analogue
               imaginal (visual, kinesthetic, or auditory)
                 static
                 dynamic
    These modes differ in:
               the transformations which can be applied
               to them (the constraints on these
               representations can be "strict" or "elastic");
               the mental work load they impose on the
               information processing capacity

(2) Available procedures
    Primarily context specific are:
      (i) rules (e.g. grammars, arithmetics, "Gestalt laws" in perception, etc.) which interact with the above-mentioned applicable transformations
    Less context specific are:
      (ii) heuristics, which can be regarded as tools for narrowing down the number of possible candidates among the transformation (e.g. concentrate on confirmatory evidence)
    Least context specific and therefore applicable in situations of information overload are:
      (iii) rules of thumb, which can be applied almost without any constraints. They provide "quick and dirty" procedures for the reduction of information so that the processing overload is relieved (e.g. "do whatever comes to your mind first")

---

game", that is, they assume that conversational conventions also hold for syllogistic reasoning. On the other hand, rules for an argumentative discourse do not presuppose a precise meaning of the propositions used, but the rules for an argumentative discourse (e.g. as stated in the Gricean maxims) are not restrictive enough to analyze them formally.

For these reasons the approach taken here models the meaning of propositions as possibility functions in fuzzy set theory. The possibility functions for the meanings of concepts can be determined experimentally; that is, they are modelled according to the actual usage in normal language. According to fuzzy set theory the meaning of concept "x" in a universe of discourse "U" can be modelled by the possibility function for "x" in U, which indicates for which states in U the concept "x" fits, for which it

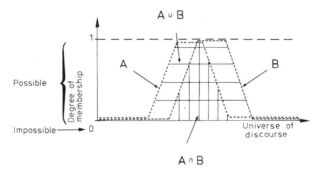

FIG. 1. Possibility distributions and binary operations upon them.

can be possibly applied, and for which it does not fit at all. Figure 1 depicts the possibility functions for two concepts as well as the basic two binary operations:

(i) the intersection: $A \cap B = \min_U (f_A; f_B)$

and

(ii) the disjunction: $A \cup B = \max_U (f_A; f_B)$.

In order to circumvent the problems in determining exact individual possibility functions [see Zimmer (1980, 1982, 1983); but Hersh & Caramazza (1976); or, for theoretical analyses, Kaufmann (1975) and Dubois & Prade (1980)] for the data reported here the possibility functions are restricted to the values: "absolutely possible" (poss $(x) =$ 1); "possible" ($0 <$ poss $(x) < 1$), and "impossible" (poss $(x) = 0$). Freksa (1981, 1982) has shown that for practical purposes this amount of specification is sufficient. Furthermore, in this approach one can avoid the debated question of how exact the information about one's own knowledge can be (cf. Nisbett & Wilson, 1977) without losing the amount of specificity necessary to apply fuzzy set theory to verbal concepts. The meaning of these concepts is then given by the elastic constraints imposed upon them by the possibility functions.

Begg (1982) concluded from his experiments that in reasoning, people apply what they are most familiar with: the rules underlying conversation and language. Following this argument, it is necessary to model expert reasoning according to the rules of argumentation in discourse. One linguistic means for expressing these rules are·quantifiers (e.g. "all", "some", "none", and "not all"). Goguen (1969) has proposed a seemingly straightforward method for modelling the meaning of vague quantifiers in this framework: starting from the meaning of crisp quantifiers in classical logic he fuzzifies them by changing the crisp constraints into elastic ones. However, one can easily see that the range of situations to which these quantifiers can be applied differs markedly: whereas "all" and "none" fit only to a very restricted range, "some" and "not all" are applicable to nearly the full range (see Fig. 2). Starting from a language-pragmatic point of view, a different approach to the modelling of vague quantifiers

FIG. 2. Scope diagrams for fuzzy quantifiers according to Goguen (1969).

for normal language seems to be more plausible: if it is assumed that the meaning of words emerges under the constraint that communicability is optimized, then the meaning of quantifiers should be modelled by possibility functions of about the same shape and the same range of applicability (see Fig. 3), as argued in Zimmer (1980). Zimmer (1982) has analyzed experimental data for this model. It turned out that, despite high intersubjective consistency, the data could not be fitted to the assumed possibility functions of Fig. 3. By analyzing the content of the items used for the empirical determination of the interpretation of the quantifiers it was found that the interpretation of the quantifiers changed with the context: for instance, in the context of science the quantifier "all" had a more restricted interpretation than in the context of everyday events. The influence of context again can be modelled by fuzzy set theory. If one determines the possibility functions for statements in the contexts (e.g. "How

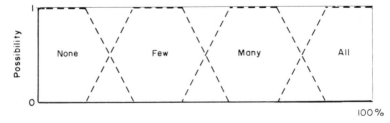

FIG. 3. Assumed scope functions for normal-language quantifiers ("all", "many", "few", and "none").

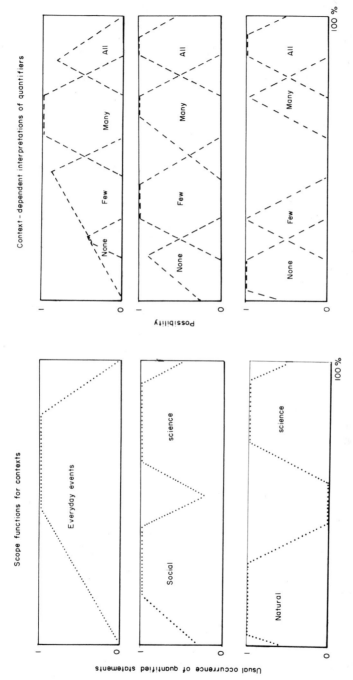

FIG. 4. The context-specific meanings of quantifiers in the contexts "science", "humanities", and "everyday events".

often does it happen in context 'A' that a statement is made, which is true in approximately $x\%$ of all cases?'') then the context—specific interpretation of a quantifier is given by its context-independent possibility function modified by the given context. The modification is done by the MIN-rule for the "and" operator. The data fit quite well to this simple model for the context-specific interpretation of quantifiers (see Fig. 4 for illustration).

In the model developed so far, the representation of propositional knowledge can be interpreted numerically by means of the elastic constraints imposed on the interpretations. In order to give a formal description of human predictive behavior it is necessary to define additionally how new information is integrated into the existing knowledge.

A model for integrating new information into knowledge

Various experiments on the capacity of humans to update their existing knowledge in the light of new information have revealed that people tend to stick to their initial opinion very tenaciously (see Phillips & Edwards, 1966). This bias towards suboptimal information processing has been termed conservatism; yet in one experiment Zimmer (1983) was able to show that it is not information in general which is processed conservatively, but predominantly the numerically expressed information. For this reason, a more general approach to the modification of knowledge in the light of new information is taken here. It starts from the assumption that the existing knowledge is represented in verbally stated propositions consisting of vague concepts and vague quantifiers. It is furthermore assumed that the resistance of propositions against modification depends on the time this knowledge has remained the same and/or on the amount of supporting information amassed for the proposition in question. The impact of the new information then depends on the salience this information has, compared with the resistance of the existing knowledge against change. These assumptions can be modelled as operations on fuzzy sets, which modify the possibility functions accordingly.

The resistance of a given quantified proposition, Q at time $t(i)$, is

$$b_i = f[d(\mathbf{Q}_i; \mathbf{Q}_{i-1})], \qquad 0 \le b_i \le 1, \tag{1}$$

where f is a monotonically decreasing function and d is the fuzzy distance (see Kaufmann, 1975). The saliency of new information is given by α ($0 \le \alpha \le 1$). The integration of new information, $I(i+1)$, into the existing knowledge, $Q(i)$, is then modelled by

$$\mathbf{Q}_{i+1}^{(x)} = \max_{x} \left[b\mathbf{Q}_i^{(x)} + (1-b)\mathbf{I}_{i+1}^{(x)}; \frac{\min_{x}[\alpha\mathbf{I}_{i+1}^{(x)}; \mathbf{Q}_i^{(x)}]}{\max_{x}[\mathbf{I}_{i+1}^{(x)}; \mathbf{Q}_i^{(x)}]} \right]. \tag{2}$$

This model for the revision of world knowledge cannot be tested directly. One possible way would be to ask subjects for their subjective estimates of the impact of old and new information. For the reasons stated above, this does not seem to be a viable approach because it would force subjects to assign numbers to ingredients of their knowledge which are most probably represented in a verbal propositional mode. The approach chosen instead consists of the following steps.

A. Determination of independent or initial (uncontrolled) variables:

(i) *calibration of fuzzy quantifiers.* The conversational interpretation of the following quantifiers has been determined empirically: practically all (always), many (often), few (seldom), practically none (never).

(ii) *calibration of adequacy of assessments.* How adequate the initial knowledge remains after new information has been given (e.g. "on the contrary", "definitely not", "perhaps", "indeed").

(iii) *calibration of the belief strength for the existing knowledge* (strong, intermediate, weak).

(iv) *calibration of the saliency of new information* (high, intermediate, low).

B. Determination of the dependent variable:

Observation of verbal statements about the belief strength after quantified statements have been confronted with quantified new information, which is either confirmatory, neutral, or conflicting.

The experiment was done with 15 German undergraduate students in psychology. They were asked to give examples for quantified statements (see (i)) from their own knowledge and to indicate how strong they believed that these statements were true (see (iii)). A typical example for such a statement is: "If I attend all the classes of a course, I will always succeed in the final exam" (belief strength low). Afterwards they were given new information (see (v)) which varied in saliency (see (iv)). Their verbal reactions to this new information fit well into the verbal expressions calibrated in step (ii). Every subject received 144 items of new information, that is, one item for every combination of conditions (i), (iii), and (iv). The results of the verbal reactions of a single subject to new information is given in Table 2; the belief strength in the existing knowledge and the saliency of the new information were both intermediate.

The regularity of the entries in Table 2 indicates that this subject processed the information in a systematic fashion and that she took the new information into account.

TABLE 2

Old knowledge $\mathbf{Q}_i$	Quantified new information $(\mathbf{I}_{i+1})$			
	Practically all (always)	Many (often)	Few (seldom)	Practically none (never)
Practically none (never)	on the contrary	definitely	perhaps	indeed
Few (seldom)	definitely not	perhaps	indeed	perhaps
Many (often)	perhaps	indeed	perhaps	definitely not
Practically all (always)	indeed	perhaps	definitely not	on the contrary

Since all the possibility functions for the quantifiers and the assessments in the table are known, it can be determined whether equation (2) models the revision of world knowledge adequately. The fit of a model is about equally high for the conditions where the belief in the existing knowledge is either intermediate or low. In the case of strong belief in the existing information the model predicts a stronger change than the one actually occurring.

With the model developed so far, it is possible to predict the impact new information will have on an existing body of knowledge. It can also be used to analyze the initial belief strength after the impact of the new information has been observed. Qualitative arguments can therefore be described by numerically stated elastic constraints, which in turn can be implemented into a numerical forecasting system.

Up to this point only the modification of one proposition at a time has been modelled. Usually predictions rely not only on one statement or proposition, but on a number of them usually organized by means of causal relationships. Nisbett & Ross (1980) report a couple of studies which indicate that humans usually organize their world knowledge by causal relations, even if the information given is not causal, but merely diagnostic. This preferred mode of knowledge organization in the form of causal schemata has to be taken into account in models of human prediction. I am working on the generalization of the approach developed above to more than one related or unrelated proposition. The idea behind this generalization is to model the subjects' assumptions about causal relationships, that is causal schemata underlying the individual knowledge base. These schemata are used to interpret qualitative arguments (Zimmer, submitted). The experiment which motivated this generalization makes clear why I assume that this approach might probably be fruitful for forecasting.

Verbal vs numerical forecasting: an application to exchange rates

A typical example for an economic prediction task of high complexity is the forecasting of exchange rates. This is a very common task for bank clerks, especially in European countries, because the commerce depends heavily on the correct timing of buying and selling (that is, whenever the exchange rates are favorable). The influences on the exchange rate stem from quantitative variables (GNP percentage increase, amount of budget deficit, interest rates, etc.) as well as from qualitative variables (stability of governments, general climate in economy, etc.).

In order to find out how experts reason in such a situation, I have performed an experiment[†] in which subjects (24 German bank clerks responsible for foreign exchange) were asked to predict what the exchange rate between the U.S. dollar and the Deutschmark would be after four weeks. Twelve subjects had to give the predictions "in their own words as if they were talking to a client", whereas the other 12 were asked to give numerical estimates (percentage of change). These two experimental conditions were chosen because the verbal predictions are what clients usually ask for, whereas the numerical predictions resemble the predictions made by the economic forecasting institutes in Germany twice a year. Both groups were asked to verbalize the steps of reasoning leading to their prediction. In order to make possible a comparison of the predictions, it was necessary to calibrate the judgments of the first group. This

---

† In a different context I have reported other aspects of the results of this experiment (Zimmer, 1983).

was done in an interactive procedure where subjects had to give verbal labels for differences in exchange rates presented to them on a CRT-monitor by a computer (TRS 80). The comparison of the verbal predictions with the estimates of the numerical forecasting group revealed that the first group was more correct and more internally consistent. While this is interesting in itself, another point might be more important: the slight difference in the instructions given caused marked differences in the way the subjects performed their task, as revealed by the verbal reports. The verbal prediction group used quantitative variables (e.g. the percentage GNP increase) as well as qualitative variables (e.g. the stability of the German government) for deriving their predictions, whereas the numerical forecasting group merely took into account variables which are usually expressed numerically. Furthermore, the verbal reports of the subjects' reasoning revealed that the verbal prediction group applied highly elaborated causal schemata, into which they fitted their assessments of the quantitative and qualitative variables. On the other hand, the protocols of the numerical prediction group consisted mostly of unconnected lists of singular assessments. From this result it seems plausible to assume that the superiority in the verbal forecasting condition is caused by the fact that the knowledge base on which these subjects relied was broader and allowed for more elaboration. However, it has to be kept in mind that the heuristic of causal schemata can be also misleading; Nisbett & Ross (1980) report ample evidence for the deleterious effects of misinterpreting diagnostic information as causal. The major difference between the studies reported in Nisbett & Ross (1980) and this experiment lies in the fact that the bank clerks were actively searching for information and only implemented their own knowledge into their reasoning.

Conclusions

The implications of the suggested model can be captured in the following way:

(i) the knowledge base for intuitive forecasts is internally represented in a propositional verbal mode; and

(ii) the reasoning underlying these forecasts is governed by communicative constraints and follows elaborated causal schemata.

The first implication can be tested by comparing the model with a numerical alternative in which numbers are assumed to be fuzzy, that is, characterized by elastic constraints in $\mathbb{R}$ (see Yager, 1983), and in which they are individually calibrated. If this alternative model describes the predictive behavior as well, one can conclude that it is not the underlying mode of knowledge representation that is decisive, but the means of handling vagueness in subjective judgments.

The studies of Begg (1982), Zimmer (1982, submitted), as well as the theoretical analysis of "rational belief" by Kyburg (1983) indicate that describing human reasoning in the framework of classical logic might be a mistaken approach. However, the suggested alternative has to become more strictly formalized in order to allow for decisive tests [see, for instance, Smith (1982) for the controversies about the "Gricean maxims" in conversation]. On the other hand, the assumption of causal schemata necessitates the investigation under which situational conditions this heuristic is applicable and under which it leads to biases.

Given that the conditions can be identified under which the suggested model captures the information in intuitive forecasts, an integrated framework for prediction becomes possible. The interaction of qualitative and quantitative aspects in forecasting can be examplified in a generalized version of "a product's lifecycle" by Chambers, Mullick & Smith (1974).

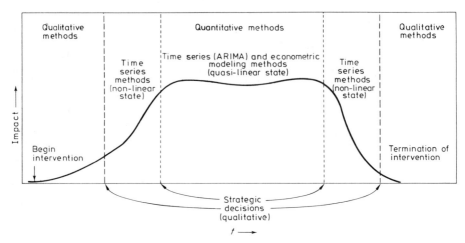

FIG. 5. The "life cycle" of an impact in its relation to forecasting techniques [modified from Chambers *et al.* (1974)].

TABLE 3

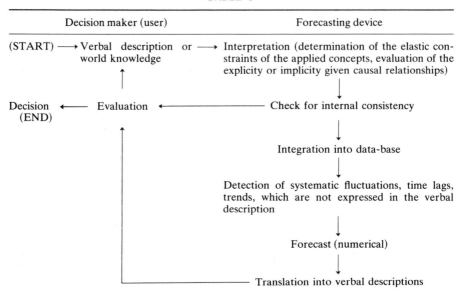

Decision maker (user)	Forecasting device
(START) ⟶ Verbal description or ⟶ world knowledge	Interpretation (determination of the elastic constraints of the applied concepts, evaluation of the explicity or implicity given causal relationships)
Decision ⟵ Evaluation ⟵ (END)	Check for internal consistency
	Integration into data-base
	Detection of systematic fluctuations, time lags, trends, which are not expressed in the verbal description
	Forecast (numerical)
	Translation into verbal descriptions

Figure 5 indicates not only the different qualitative and quantitative phases in an impact analysis but, furthermore, the qualitative strategic reasoning underlying the decisions about the applicability of different forecasting techniques.

Viewing forecasting as a means to improve decision making forces one to look for efficient ways of making the forecasting information usable for the decision maker. The approach reported here suggests an interactive forecasting process as depicted in Table 3.

In this interactive forecasting process the right-hand side consists of the forecasting expert system which determines interactively the individual knowledge base and the derived predictions. Furthermore, it simultaneously analyzes quantitative external data as well as numerical interpretations of qualitative judgments by means of traditional time-series analyses and forecasting techniques. In order to get it working it will be necessary to develop further the interpretation and translation algorithms and the evaluation of underlying causal schemata. Nevertheless, this sketch for a forecasting expert system might provide a framework for future developments, which bridge the gap between the expertise and the prediction skills on the side of the decision maker and the analytical tools of numerical forecasting.

This is an expanded version of a paper read at the Second International Symposium on Forecasting, Istanbul, 1981. I want to thank the other contributors for many valuable suggestions made in the discussion. The final form of this manuscript has gained much clarity in style as well as in arguments from discussions with Jennifer J. Freyd and Gisela Redeker.

References

BEGG, I. (1982). On the interpretation of syllogisms. *Journal of Verbal Learning & Verbal Behavior*, **21**, 595–620.

BUTLER, W. F., KAVESH, R. A. & PLATT, R. B., Eds (1974). *Methods and Techniques of Business Forecasting.* Englewood Cliffs, New Jersey: Prentice–Hall.

CHAMBERS, J. C., MULLICK, S. K. & SMITH, D. D. (1974). *An Executive Guide to Forecasting.* New York: John Wiley and Sons.

CLEARY, J. P. & LEVENBACH, H. (1982). *The Professional Forecaster: The Forecasting Process through Data Analysis.* California: Lifetime Learning Publications.

DUBOIS, D. & PRADE, H. (1980). *Fuzzy Sets and Systems: Theory and Application.* New York: Academic Press.

EDWARDS, W. (1968). Conservatisms in human information processing. In KLEINMUNTZ, B., Ed., *Formal Representation of Human Judgment*, pp. 17–52. New York: Wiley.

FRESKA, C. (1981). Linguistic pattern recognition. *Ph.D. thesis*, University of California at Berkeley.

FRESKA, C. (1982). Linguistic description of human judgments in expert systems, and in "soft" sciences. In TRONCALE, L., Ed., *Systems Science and Science.* Louisville, Kentucky: Society for General Systems Research.

FRIEDMAN, M. I. & WILLIS, M. R. (1981). *Human Nature and Predictability.* Lexington, Massachusetts: Lexington Books.

GOGUEN, J. A. (1969). The logic of inexact concepts. *Synthèse*, **19**, 325–373.

HERSH, H. M. & CARAMAZZA, A. (1976). A fuzzy set approach to modifiers and vagueness in natural language. *Journal of Experimental Psychology, General*, **105**, 254–276.

HOGARTH, R. M. & MAKRIDAKIS, S. (1981). Forecasting and planning: an evaluation. *Management Science*, **27**(2), 115–138.

KAHNEMAN, D. & TVERSKY, A. (1973). On the psychology of prediction. *Psychological Review*, **80**, 237–251.

KAHNEMAN, D., SLOVIC, P. & TVERSKY, A., Eds (1982). *Judgment under Uncertainty: Heuristics and Biases.* Cambridge: Cambridge University Press.

KAUFMANN, A. (1975). *Theory of Fuzzy Subsets*, vol. 1. New York: Academic Press.

KYBURG, H. E. (1983). Rational belief. *The Behavioral and Brain Sciences*, **6**(2), 231–273.

LENAT, D. B. (1982). The nature of heuristics. *Artificial Intelligence*, **19**, 189–249.

LINDZEY, G. (1965). Seer versus sign. *Journal of Experimental Research in Personality*, **1**, 17–26.

MEEHL, P. E. (1954). *Clinical versus Statistical Prediction.* Minneapolis: University of Minnesota Press.

MURPHY, A. H. & WINKLER, R. L. (1975). Subjective probability forecasting: some real world experiments. In WENDT, D. & VLEK, C., Eds, *Utility, Probability, and Human Decision Making.* Dordrecht: Reidel.

NISBETT, R. E. & ROSS, L. (1980). *Human Inference: Straategies and Shortcomings of Social Judgement.* Englewood Cliffs, New Jersey: Prentice-Hall.

NISBETT, R. E. & WILSON, T. D. (1977). Telling more than we know: verbal reports on mental processes. *Psychological Review*, **84**, 231–259.

PHELPS, R. H. & SHANTEAU, J. (1978). Livestock judges: how much information can an expert use? *Organizational Behavior and Human Performance*, **21**, 204–219.

PHILLIPS, L. & EDWARDS, W. (1966). Conservatism in a simple probability inference task. *Journal of Experimental Psychology*, **72**, 346–354.

SHORTLIFFE, E. H. (1976). *Computer-based Medical Consultations: MYCIN.* New York: Elsevier.

SJÖBERG, L. (1981). Aided and unaided decision making: improving intuitive judgement. *Journal of Forecasting*, **1**, 349–363.

SMITH, N. V., Ed. (1982). *Mutual Knowledge.* London: Academic Press.

YAGER, R. R. (1983). An introduction to applications of possibility theory. *Human Systems Management*, **3**, 246–253.

ZIMMER, A. C. (1980). Eine Formalisierung mnestisch stabilisierter Bezugssysteme auf der Grundlage von Toleranzmengen. In THOMAS, A. & BRACKHANE, R., Eds, *Wahrnehmen–Urteilen–Handeln.* Bern: Huber.

ZIMMER, A. C. (1982). Some experiments concerning the fuzzy meaning of logical quantifiers. In TRONCALE, L., Ed., *Systems Science and Science.* Louisville, Kentucky: Society for General Systems Research.

ZIMMER, A. C. (1983). Verbal vs. numerical processing. In SCHOLZ, R., Ed., *Individual Decision Making under Uncertainty.* Amsterdam: North-Holland.

ZIMMER, A. C. (submitted). A model for schema guided reasoning.

Index

DATE DUE